797,885 Books
are available to read at

www.ForgottenBooks.com

Forgotten Books' App
Available for mobile, tablet & eReader

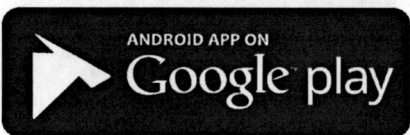

ISBN 978-1-332-41933-3
PIBN 10424639

This book is a reproduction of an important historical work. Forgotten Books uses state-of-the-art technology to digitally reconstruct the work, preserving the original format whilst repairing imperfections present in the aged copy. In rare cases, an imperfection in the original, such as a blemish or missing page, may be replicated in our edition. We do, however, repair the vast majority of imperfections successfully; any imperfections that remain are intentionally left to preserve the state of such historical works.

Forgotten Books is a registered trademark of FB &c Ltd.
Copyright © 2015 FB &c Ltd.
FB &c Ltd, Dalton House, 60 Windsor Avenue, London, SW19 2RR.
Company number 08720141. Registered in England and Wales.

For support please visit www.forgottenbooks.com

1 MONTH OF FREE READING

at

www.ForgottenBooks.com

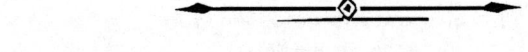

By purchasing this book you are eligible for one month membership to ForgottenBooks.com, giving you unlimited access to our entire collection of over 700,000 titles via our web site and mobile apps.

To claim your free month visit: www.forgottenbooks.com/free424639

* Offer is valid for 45 days from date of purchase. Terms and conditions apply.

Similar Books Are Available from
www.forgottenbooks.com

General Biology
by Leonas Lancelot Burlingame

A Manual of Bacteriology
by Herbert U. Williams

The Human Body
A Popular Account of the Functions of the Human Body, by Andrew Wilson

Practical Biology
by W. M. Smallwood

Manual of Practical Anatomy, Vol. 1
Upper Limb; Lower Limb; Abdomen, by D. J. Cunningham

Problems of Biology
by George Sandeman

Science and Human Affairs from the Viewpoint of Biology
by Winterton C. Curtis

History of Biology
by L. C. Miall

The Bacteria
by Antoine Magnin

A Manual and Atlas of Dissection
by Simon Menno Yutzy

Practical Microscopy
by Maurice Norton Miller

Elements of the Comparative Anatomy of Vertebrates
by W. Newton Parker

The Elements of Animal Biology
by S. J. Holmes

Handbook of Plant Dissection
by J. C. Arthur

Biology and Its Makers
With Portraits and Other Illustrations, by William A. Locy

An Introduction to Vertebrate Embryology, Based on the Study of the Frog, Chick, and Mammal
by A. M. Reese

Elementary Biology
Plant, Animal, Human, by James Edward Peabody

Scientific Method in Biology
by Elizabeth Blackwell

The Biology of Birds
by J. Arthur Thomson

Fathers of Biology
by Charles McRae

KETCHUM'S LESSONS ON THE EYE

Dedicated to the "World of Optometry" and
Especially to Those Who Have Sacrificed
Their Time and Energy to the End
That the Word "Optometrist"
May Be Honored by All
Other Professions

THE CONTENTS OF THIS BOOK HAS BEEN ESPECIALLY ARRANGED TO MEET THE PRESENT DAY NEEDS OF THE OPTOMETRY STUDENT. IT COVERS THE ESSENTIALS OF THE "STRUCTURE AND FUNCTION OF THE EYE, THE ORBIT AND ITS APPENDAGES". ALSO THE DISEASES OF THE EYE THAT HE SHOULD RECOGNIZE

Edited and Published by

MARSHALL B. KETCHUM, M. D.

President and Founder of the
Los Angeles Medical School of Ophthalmology and Optometry.

Los Angeles, California.

1920

MARSHALL B. KETCHUM, M. D.

A FEW APHORISMS

OPTOMETRY, TODAY, IS THE HIGHEST CONCEPTION THAT THOSE BEST QUALIFIED TO KNOW, THINK IT IS.

WE CANNOT HARMONIZE OUR TEACHING WITH THE STUDENT'S CAPACITY FOR LEARNING, NOR HIS IDEAS OF HIS SPECIAL REQUIREMENTS.

DO NOT BE IGNORANT OF YOUR IGNORANCE, BUT KNOW THAT YOU KNOW.

THE KNOWLEDGE OF OPTOMETRY MUST COME TO THE STUDENT IN AN ORDERLY WAY.

ALL THINGS ARE DONE BY THOUGHT.

THE DIFFERENCE BETWEEN THE OLD-TIME REFRACTING OPTICIAN AND THE QUALIFIED UP-TO-DATE OPTOMETRIST IS EXACTNESS IN DETAIL BASED UPON A DEEP KNOWLEDGE OF HIS UNDERTAKING.

ANATOMY—means **STRUCTURE**—all or any part or the body.

PHYSIOLOGY
PHYSIOLOGICAL } means **FUNCTION**. The natural action of any part of the anatomy of the body either alone or in conjunction with any other part of the body.

PATHOLOGY
PATHOLOGICAL } Pertains to a diseased condition of any part of the body.

ETIOLOGY— **CAUSE** of any fault.

DIAGNOSIS— Recognition of the nature of diseased conditions.

PROGNOSIS— Judgment formed regarding the future outcome of a diseased condition.

REFERENCE BOOKS ON GENERAL ANATOMY AND PHYSIOLOGY

CUNNINGHAM——MORRIS——GRAY

ALSO STEWART'S MANUAL OF PHYSIOLOGY

MAY ON DISEASES OF THE EYE

GOULD'S POCKET MEDICAL DICTIONARY

DEAVER'S HEAD AND NECK.

INTRODUCTORY REMARKS.

An Optometrist:—means, at the present time, anyone, who having properly qualified for the purpose, has his practice limited to the diagnosis and correction of defective vision not due to disease, as well as abnormal conditions of the muscles of the eyes amenable to relief or treatment with lenses, prisms and ocular calisthenics. Also the diagnosis of any pathological condition of the eye that he may refer the case to the Oculist for medical or surgical treatment.

Owing to unfamiliarity with the technical terms of regular medical books on the subject of anatomy, physiology and pathology of the eye and the limitations of his professional and legal requirements on these subjects, the student and practitioner of Optometry has, as a class, been unable to "get the story" as he should know it.

The Author's extended experience as an Oculist, Professor of Ophthalomology in a Medical School for several years, as well as personally training the Optometrist from the "raw into the finished product" has given him a rare insight into the exact needs of the non-medical refractionist along these special lines.

This book is edited especially for him and is entirely different from anything in print. It is not only for the student who is just entering this special field of endeavor, but for the licensed Optometrist, as well; and we might add, that the medical refractionist will find many features of interest to him.

The standpoint from which the subject matter is presented, is that of a fairly complete working knowledge of the eye and its appendages and is of sufficient scope to practically cover all that the Optometrist will be required to know of this branch for years to come.

In no sense is the minute anatomy of the eye any necessary part of an Optometrist's requirements. He does not cut, he does not treat, as that is the legal function of the Ophthalmic surgeon, but his knowledge of the eye is required to be of such a nature that for the good of the patient as well as himself and his professional associates he will be enabled to follow his calling intelligently and know his limitations.

Three main things confront all refractionists when a patient comes for advice. First, does he need glasses only? Second, does he need treatment as well as glasses, and third, does he need treatment only? This the Optometrist must know and then act accordingly. Much more might be said, but we leave that to the student as we feel sure that he will find herein much of interest and value to him.

I have planned, in my endeavor to help the student, to make this usually dry subject a decidedly interesting one to him. First, by presenting in picture form, with notes, a general scheme in their proper order, of the principal parts of the subject that he should master. Then following with a talk on each part sufficiently explicit so that a gross knowledge is methodically and easily gained. The quiz following each lesson covers the essentials and causes the student to carefully review that lesson and learn to formulate his own answers. Having completed the general outline of the work in this manner I have presented in Parts Two and Three under special headings a more intimate and complete consideration of the individual parts of the eye that was covered by the lessons, as well as many other features.

MARSHALL B. KETCHUM, M. D.

LESSON ONE.

(Section One.)

In taking up the study of the eye it is well for the Optometry student to realize that, while as a class the human eye is pretty much the same in all people, a great deal of comparative study has been necessary in order to arrive at a definite standard of anatomical and optical measurements and the principles therein involved, so as to form a basis for consideration from all standpoints; hence, the term "SCHEMATIC EYE" has been adopted to cover what may be considered as the perfect eye.

The following figures are given to convey at once a gross conception of the eyes and the orbits, along with the general scheme of their relationship, and this will lead to an interest in the text that follows.

THE METRIC SYSTEM.

Everyone should be familiar with the metric system of measurements.

The inch, foot, and yard system is practically obsolete in anything but gross work, though the average person does not seem to know it. A fiftieth or a thousandth of an inch or yard is practically an indefinite quantity and the scientific man has no use for it. Such a system should be entirely abolished as being too crude for the present day requirements. "Decimals and fractions" is the only definite, certain, and easy way of obtaining or properly explaining any weight or measure that requires delicate consideration. The importance of possessing a uniform system of measures that is subject to infinite and exact consideration has been recognized by scientists generally for some time past. In Europe it is practically the only method in use. It is called

The Metric System and in calculation corresponds with the way we figure dollars and cents. It is founded on the word **metre,** which is the unit of length, based on the measurement of the quadrant of a meridian of the earth. There are **only three parts** of this system that we usually use in our ordinary ophthalmic and optical measurements. They are—Metre, Centi-metre, Milli-metre. Study this table a few minutes and you will have it. The Optical houses furnish small ivory rules and cards showing by measure exactly, centi-metres and milli-metres as compared with the inch system.

Metre	Centi-metres	Milli-metres
1 =	100 =	1000

Explanation—The smallest decimal we use is 1 millimetre (mm.) and it takes 10 mm. to make 1 centimetre (cm.) and that's all there is to it. Compared with the inch system, the optical student should be able to transpose from one to the other. It is as follows: 1 m. = about 40 inches (1 yard and 4 inches.) 25 mm. = about 1 inch. If 10 mm. = 1 cm., then it would take 2:5 cm. to make 25 mm. or 1 inch.

In all optical problems these relative measures are used. Relative to the dollar system, it is like this: 1 meter = 1 dollar; 1 centimetre = 1 cent; 1 mm. = 1/10 of a cent. Now looking at a rule with millimetres and centimetres marked off on it you will become familiar by sight just what each one is as to distance. Here it is.

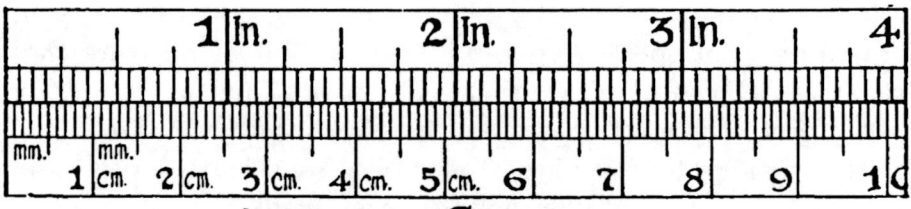

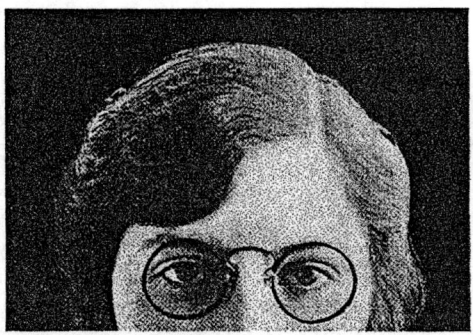

Figure 1.

Every normal person has two eyes, (either, being a perfect mate for its fellow eye) so situated in their respective positions in the head as to be parallel one with the other and under such control of the eye muscles as to work and move together in perfect relationship when looking at objects at any and all distances.

They are located in the upper and front part of the skull, each in a bony cavity commonly called the orbit (eye socket).

The distance between the two eyes varies somewhat in different individuals, owing to the fact that the general measurements of any two heads are not identical. However, in the average adult, we usually find that from the center of one pupil to the center of the pupil of the other eye, (called pupillary distance—abbreviated P. D .) it is about 60 millimetres, while the average range of distance between the eyes is from 56 mm. (2¼ inches) to 60 mm. (2½ inches).

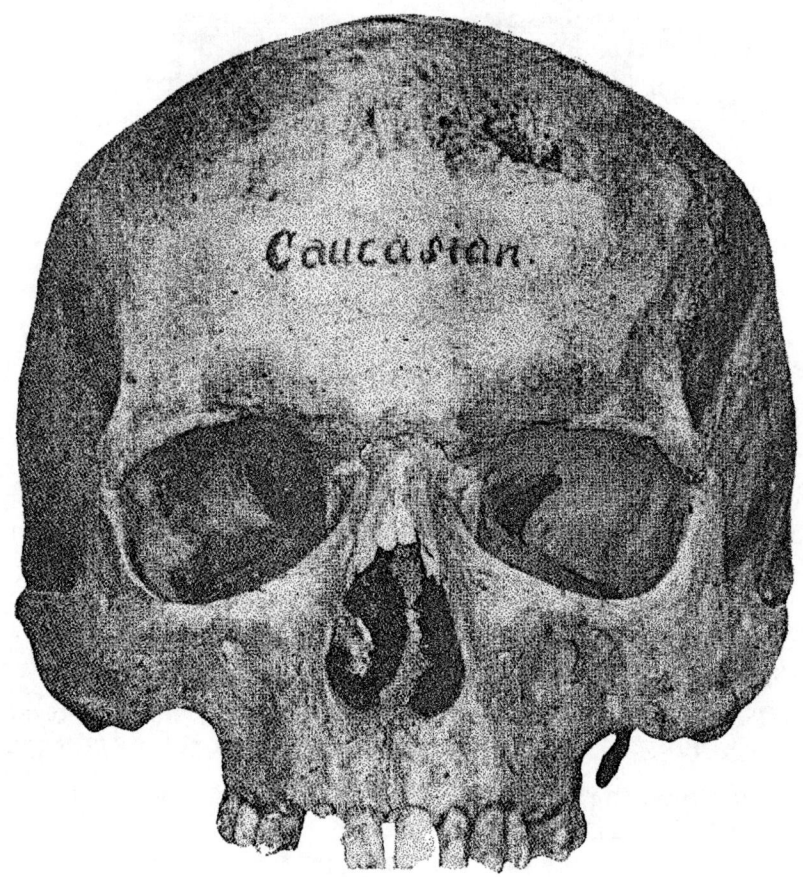

Figure 2

Figure 2 is to illustrate the fact that although the two eyes are parallel the two orbits are **not** parallel, but diverge from one another; also that in front (the base) they are somewhat oblong and irregular in form and droop downward.

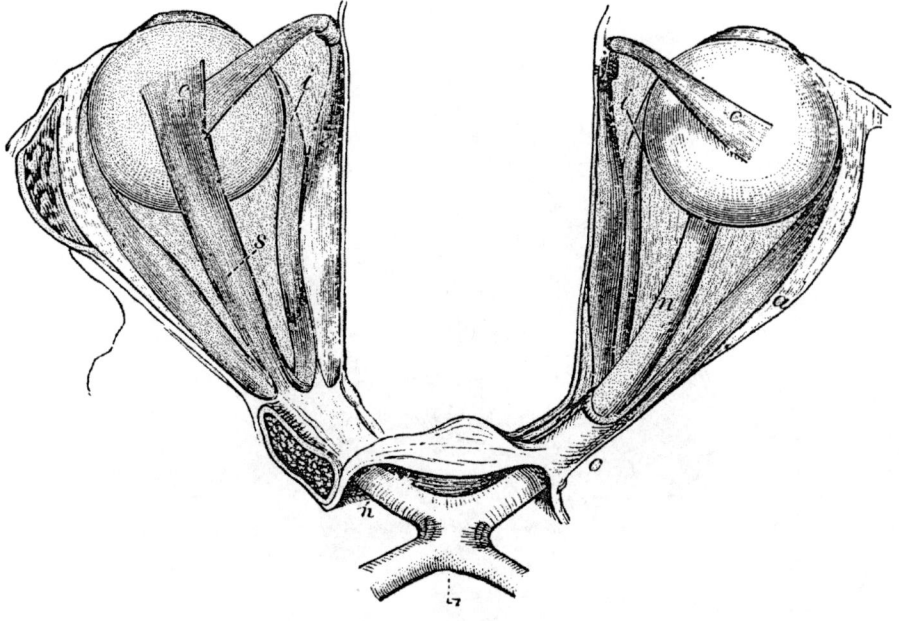

Figure 3

Figure 3 is to show the position of the orbits and how the eyes lie in each orbit and are held in a position of parallelism by the muscles that control their action. See also how they are connected with one another where the optic nerves join, inside of the skull, and connect with the brain.

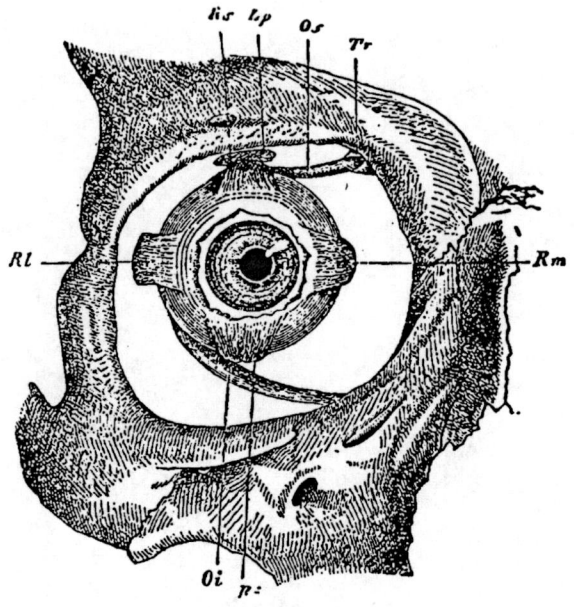

Figure 4

Figure 4 shows front view of the right eye in the orbit and how the muscles hold it in its proper position straight ahead, and these muscles are so arranged that the eye can be moved in any direction with the slightest effort.

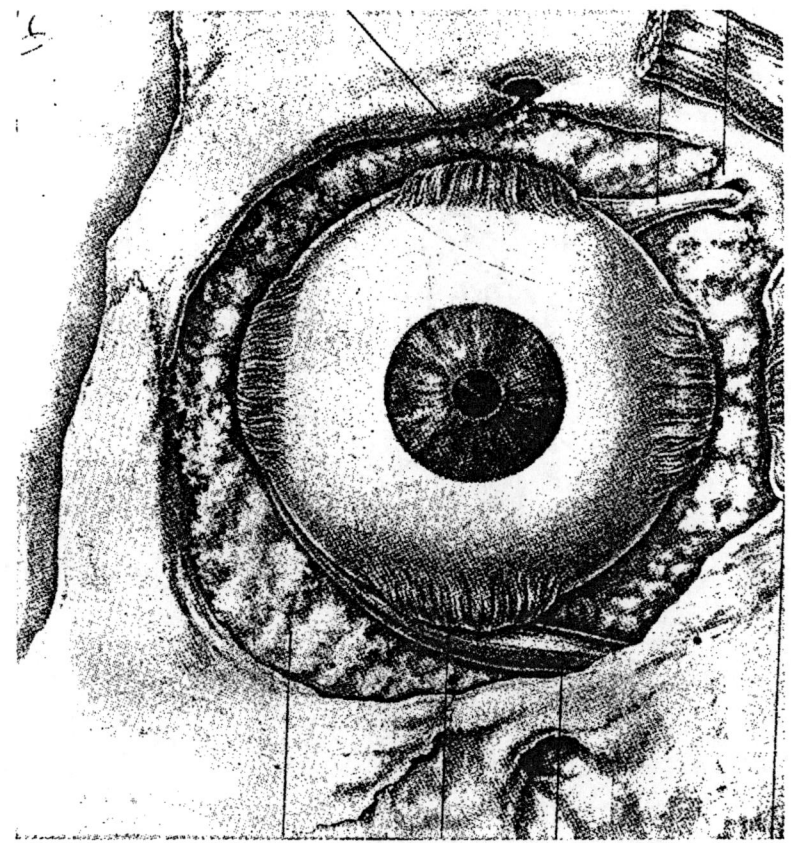

Figure 5.

Figure 5 shows the eye surrounded by a soft cushion of fat which offers no resistance to its movements in any direction. It also acts as a support to the eye.

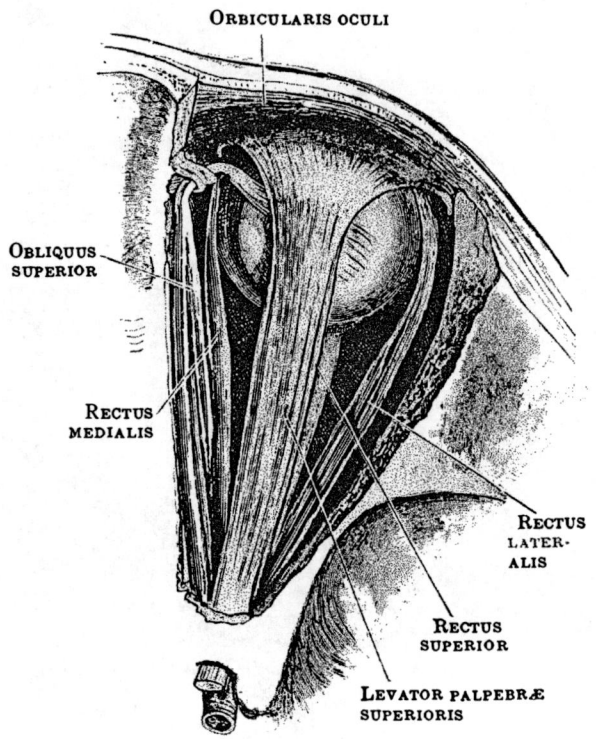

Figure 6

Figure 6 shows a view of the right eye from above—top of skull removed. The position of the eye-ball within the orbit along with the arrangement of some of the muscles that control its movements.

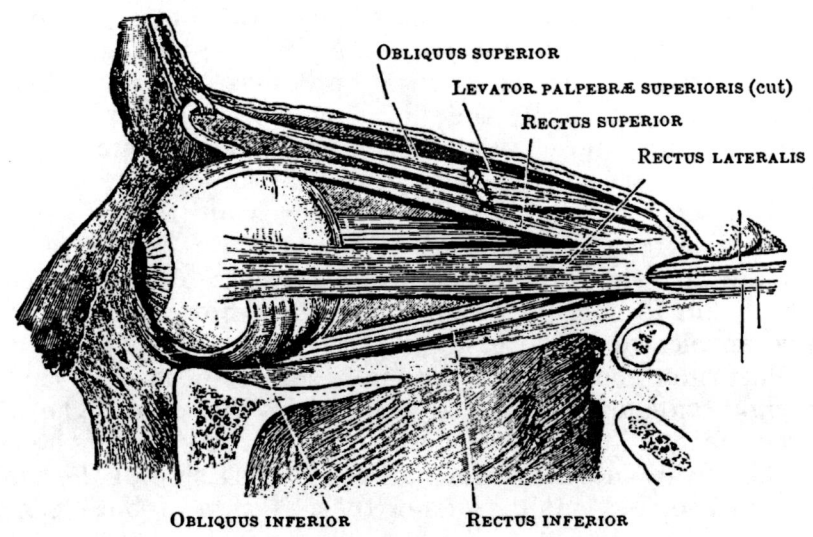

Figure 7

Figure 7 shows side view of eye and muscles which gives an idea of how the eye-ball lies within the orbit.

The act of seeing distinctly is a peculiar one and involves many fine points for consideration. Figure 8 is a sketch made to illustrate the fact that the eye alone is practically only a medium through which vibrations of light coming from some object outside of the eye are properly adjusted within it. These visual sensations are then conducted on through the optic nerve, to where it joins the optic nerve of the fellow eye at the optic chaism (O.C.), where all points are fused together, and from there on through the right and left optic tracts to different lobes of the brain where the "sense of sight" is located. The diagram shows how the corresponding sides of the retina are united—at the optic chaism—by crossed fibres and accounts for the "field of vision" in each eye and that objects seen with both eyes are united where the optic nerves meet, the fibres on the right side of both nerves uniting there, and after union going thence into the brain in the nerve which is on the right side of the head, and the fibres on the left side of both nerves uniting in the same place, and after union going into the brain in the nerve which is on the left side of the head, and these two nerves meeting in the brain in such a manner that their fibres make but one entire species of picture, half of which in the right side of the sensorium comes from the right side of both eyes through the right side of both optic nerves to the place where the nerves meet, and from thence on the right side of the head into the brain, and the other half on the left side of the sensorium comes in like manner from the left side of both eyes.

HEREDITY.

The influence of heredity on the eye and its appendages is particularly noticeable in a great many families, as recent studies show more and more the tendency of the offspring in many ways to resemble the parents even in the most minute details of structure and this fact deserves careful consideration as this subject has not been given the critical study that it deserves.

At birth the two human eyes do not work in perfect harmony together. Meaning that the influence of the action of the muscles that hold each one in position, is not under any kind of control, physiologically speaking, at this time; so that the infant may, very early in life, look more or less cross-eyed until the necessity for binocular fixation comes into play. In the course of from six months to a year, as the eyesight devel-

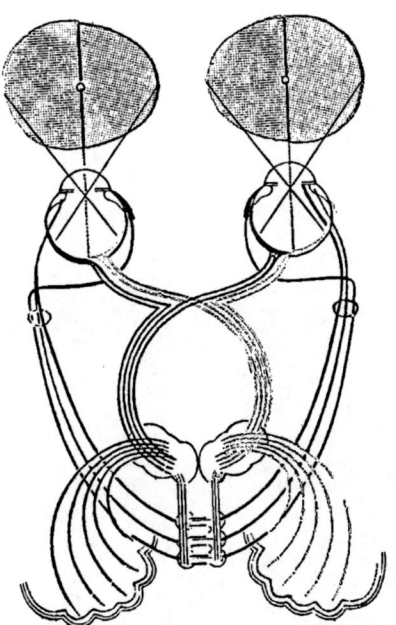

Figure 8

For description see page 20.

opes along with the slowly growing intellect of the child it begins to take notice of moving objects and directs its eyes toward them, and thus in the course of time he learns by experience to judge the distance and locality of an object as well as its physical characteristics. Seeing the same object with both eyes at the same time demands certain adjustments of the position of each eye in relation to one another in order that at all times and all positions and distances by **perfect fusion** the two images become as one. This is a physiological function which when fully developed is called the **Fusion Power.** As a rule this fusion power is not completely developed until the end of the sixth year.

LESSON ONE.

(Section Two.)

THE ORBIT.

The orbit is the first to be considered because it serves as a protection for the eye which is imbeded within it in a body of fat, this fat, however, offers no resistance to the movement of the eye in any direction.

It is rather cone shaped with its base forward and the apex extending backward and inward at an angle of about 40 degrees with the median plane, straight ahead.

Its average dimensions are

>Depth about 45 millimeters.
>Horizontal (at base) about 40 mm.
>Vertical (at base) about 35 mm.

It has four sides, four angles and nine openings, each opening is called a foramen; plural, foramina.

Each orbit consists of seven bones which are located as follows·

>Roof (above)   1.   Frontal.
>
>Floor (below)   2.   Superior Maxillary.
>   3.   Palate.
>
>Nasal side   4.   Ethmoid.
>   5.   Lachrymal.
>
>Temporal side 6.   Sphenoid.
>   7.   Malar.

As three of these bones serve to form a part of each orbit, viz.: the frontal, ethmoid and sphenoid, it requires only eleven separate bones to form both orbits. Not all of each bone is required to form the orbit, but only what is called the "orbital portion" or part of these bones, otherwise they go to make up the skull.

The walls of the orbit as a whole form a strong, bony, ring, at its base, called the **Orbital Margin.**

The four boundaries or walls make four angles, viz.:

> Superior External Angle.
> Superior Internal Angle.
> Inferior External Angle.
> Inferior Internal Angle.

The following are the nine Foramina, viz.:

1. Supra-orbital,
2. Infra orbital,
3. Anterior-ethmoidal,
4. Posterior-ethmoidal,
5. Malar,
6. Nasal canal or groove,
7. Spheno-maxillary fissure,
8. Optic foramen
9. Sphenoidal fissure.

Of these nine openings **only two** concern the Optometrist to any extent and those are No. 8 and No. 9, because of the nerve and blood supply that enter the orbit through them.

The Optic Foramen is a small, round opening at the back part or apex of the orbit, through which the optic nerve and the ophthalmic artery enter the orbital cavity from the inside of the skull, while the

Sphenoidal Fissure, a much larger opening on the **temporal side** of the orbit serves as a passageway for the nerves, arteries and veins, viz.: The third, fourth and sixth cranial nerves; the frontal lachrymal and nasal branches of the ophthalmic or first division of the fifth nerve, branches of the sympathetic nerve, the ophthalmic veins and also lachrymal meningeal arteries.

The BLOOD SUPPLY of the eye comes from the **ophthalmic artery,** (a branch of the internal carotid). See **Part Two** for particulars.

QUIZ ON LESSON ONE (Section One).

1. What is meant by the term "schematic eye"?
2. Give average P. D. measurement. Use metric system.
3. What holds the eyes parallel with one another?
4. Are the two orbits parallel, divergent or convergent with one another in front?
5. How and where are the two eyes connected with one another?
6. What is the cushion of fat in the orbital cavity for?
7. Are the optic tracts in the brain or in the eye?
8. Study the diagram showing the working of the field of vision of both eyes together
9. At what period of life is the fusion power fully developed?

QUIZ ON LESSON ONE (Section Two).

10. Describe shape, position, and dimensions of the orbit.
11. How many degrees do the orbits diverge from the median line?
12. How many bones comprise each orbit?
13. How many bones are required to form both orbits?
14. Name and locate the seven bones.
15. Why is it that only eleven bones comprise both orbits?
16. Name the four orbital angles.
17. How many foramina in the orbit?
18. Describe the optic foramen and the sphenoidal fissure and state why they are especially mentioned.
19. What artery supplies blood to the orbital cavity?

Make your answers both oral and written. Written answers are much the best as then you can refer to the text to see if you are right.

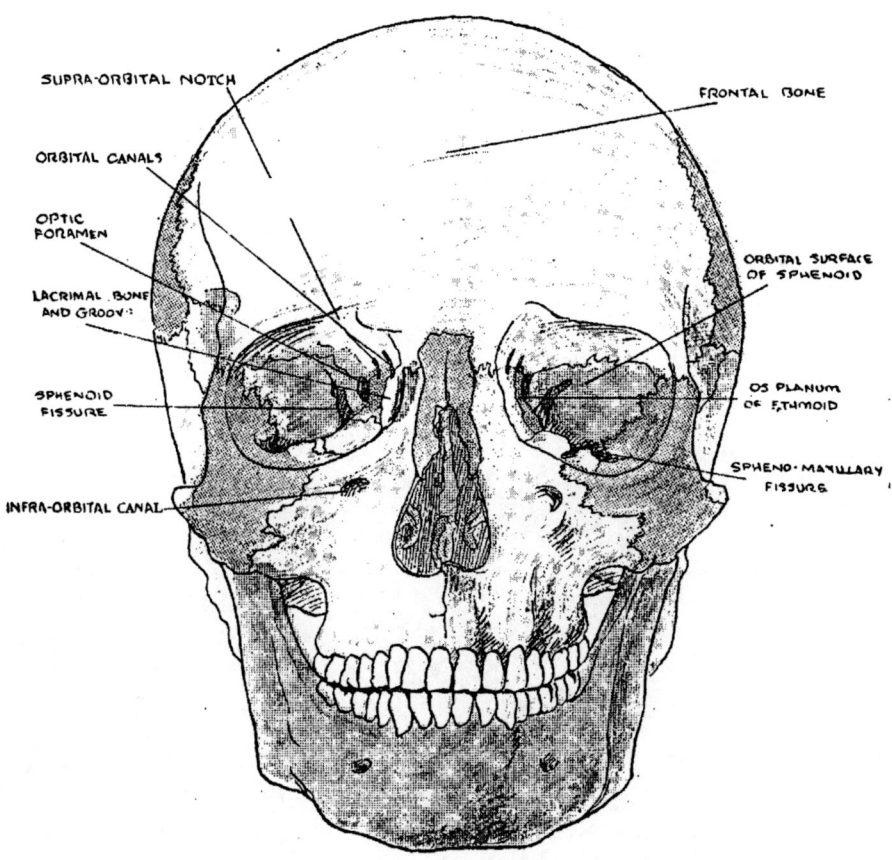

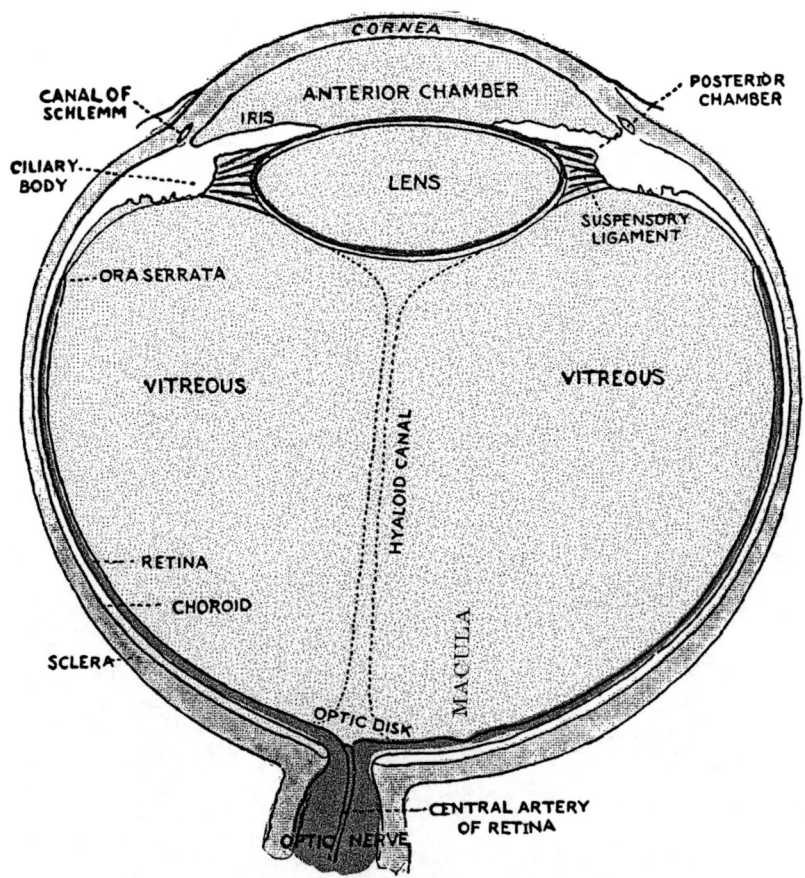

Horizontal Section of the Eyeball. Magnified about 3½ ×.

The student must realize that this picture shows a half section cut through the center of the eye horizontally in order to show the inside parts, thus making the lens look oblong. Viewed from the front the lens has the form of a trial case double convex lens.

TO THE STUDENT

At this point, as you are about to study the eye-ball, muscles. eyelids etc., be sure to look at some person's eyes as you have never done before. See the lids as they open and close. Notice how the lashes turn up or down. See the size of the eye, its color, how the pupil will contract and dilate, how the muscles move the eye in different directions and the other eye follows it. As you take up each chapter in study, look at the eyes again and again for reference. In **Part Two** you will find a special reference to each part of the eye, going more into detail. Memorize **Part One** first, then go for **Part Two** good and strong.

LESSON TWO.

THE EYE-BALL.

This lesson is merely a general outline of the eye-ball and is intended to convey to the student sufficient knowledge to serve as a working basis for that which is most essential to him at this stage of his work, so that he can at once proceed to take up the study of Ophthalmic Optics and have a clear conception of what he is doing.

The eye-ball at birth is small and nearly round, although it varies considerably in size as well as in form. Its average antero-posterior diameter is 17.3 mm., which is much less than when it is fully developed later in life. The period of its most rapid growth is during the first years in life; this is followed by a period of slower growth, although the eye-ball steadily increases in size up to the age of puberty, and when fully developed measures about 24 mm. in its antero-posterior diameter and about 23 mm. in its transverse diameter at its equator.

It is sometimes called the **globe** because it is so nearly round, and has **three coats or tunics,** viz.:

	Name.	Part of each coat.
1.	External.	Sclerotic and Cornea.
2.	Middle.	Choroid, Ciliary Body and Iris.
3.	Internal.	Retina.

The **Sclerotic,** commonly called the "white of the eye" is a heavy fibrous opaque membrane and covers the posterior five-sixths of the ball.

The **Cornea** is the clear transparent section in front and covers the anterior one-sixth.

The **Choroid** is a thin layer of brown pigment, nerves and blood vessels and lies close to the sclerotic, lining it entirely.

The **Iris** is seen in the front part of the eye directly back of the cornea. It has an opening near its center, (usually a little to the nasal side), called the **pupil.**

The **Ciliary Body** is back of and continuous with the iris, but cannot be seen from the front. It is composed of two parts the ciliary muscle and the ciliary processes. The ciliary muscle is called the "muscle of accommodation" because it aids in adjusting the focusing power—the lens—of the eye. The entire second tunic is also called the **vascular** coat, owing to its many blood vessels; another name is the uveal tract, an old-time term given to it owing to a section of the ciliary body having somewhat the shape of a bunch of grapes.

The **Retina** is a very thin transparent membrane containing arteries and veins, and lines the choroid. It is the layer that receives the outside images and pictures them upon the brain through the **optic nerve** which enters the back part of the eye a little to the nasal side of its posterior pole or axis. The exact point upon the retina where all images are actually focused is located practically in the center of the retina and is called the **macula lutea** or **yellow spot**, (usually called the macula).

Directly in the center of the macula is the principal focal point of the eye, called the **fovea centralis** or center of focus.

Within the coats of the eye we find—see diagram—spaces filled with fluids; the crystalline lens, etc., which may be described as follows, viz.·

There are three chambers in the eye:

Anterior, Posterior, Vitreous.

The **anterior** and **posterior** chambers are between the cornea and the lens, being separated by the iris but still connected by the pupil. They are called anterior and posterior because one is in front of and the other directly back of the iris.

These two chambers, being connected, are filled with a transparent watery fluid called the **aqueous humor**.

The larger chamber, back of the lens, is the **vitreous chamber**, also called the **hyaloid cavity**. This chamber is filled with a heavy, thick, transparent body called the **vitreous humor**. It somewhat resembles the white of an egg in its consistency and serves as a support to the tunics or coats of the eye in order to keep the ball in perfect shape.

The **Crystalline Lens** is a double convex, transparent, spherical body and is a little more convex behind than in front; it is located directly back of the iris on a line with the pupil and in contact with it. Its purpose is to aid in focusing images upon the macula. It is confined in a very thin transparent membrane, the **lens capsule,** and is held in position by a delicate band, the **suspensory ligament,** which entirely surrounds it. This ligament is an extension of the **hyaloid membrane,** which is a very thin, transparent body lining the entire eyeball adjoining and separating it from the contents of the large chamber which is back of the lens.

All the clear, transparent parts of the eye, namely the cornea, aqueous humor, lens and vitreous humor are, together as a whole, called the **Refractive Media,** and it is through these transparent, refracting media that all the images are focused upon the macula.

The **Optic Nerve** is a long bundle of fibres coming from the brain through the optic foramen into the orbital cavity and enters the posterior part of the eye through the sclerotic and choroid; it then expands like a cup in all directions, forming the inner layer of the retina. In looking into the eye through the pupil we see a round, whitish spot, apparently about the size of a ten cent piece; this is where the optic nerve enters the back part of the eye and is called the **optic disc**; it is also called the "blind spot". The optic nerve connects the eyeball with the brain.

The Optic Axis and Visual Line.

An imaginary line, the central line of the globe direct through the center of the cornea and the lens to a point near the inner margin of the macula is called the **optic axis.**

There is a similar term called the visual axis, which is not the optic axis, but is the line of vision, **visual line,** and is direct from the fovea to the center of the object looked at, called the **point of fixation.**

QUIZ ON LESSON TWO.

1. What is the size of the eyeball at birth?
2. What is its size when fully developed?
3. Why is it sometimes called "the globe?"
4. How many tunics? Name them.
5. Name the different parts of each tunic.
6. How much of the globe does the sclerotic and cornea each cover?
   What color is the choroid?
   Is the pupil in the center of the iris?
9. What constitutes the ciliary body?
10. What is the ciliary muscle for?
11. What and where is the vascular coat?
12. What part of the eye receives images from the outside?
13. Name and locate the exact center of focus.
14. Name and locate the chambers of the eye and their contents.
15. Describe and locate the crystalline lens.
16. Where do we find the hyaloid membrane?
17. What constitutes the refractive media?
18. Through what foramen does the optic nerve enter the orbit?
20. How can we see the optic disc?
19. What connects the eyeball with the brain?
21. What is the difference between optic axis and visual line?
22. Be sure to master the metric system of measurements.

LESSON THREE.

THE EYE MUSCLES AND THEIR NERVE SUPPLY.

Although the orbits diverge from one another, the eyes are perfectly parallel with each other and always move in perfect unison. They are enabled to do this because each eye is under the perfect control of six muscles. Altogether there are seven muscles in the orbit; the seventh one raises the upper lid. These six muscles are called the extrinsic (meaning outside) muscles. The upper lid muscle is called the **levator palpebrarum** (meaning to lift the lid).

The Extrinsic Muscles.

1. External Rectus
2. Internal Rectus
3. Superior Rectus
4. Inferior Rectus
5. Superior Oblique
6. Inferior Oblique

(The plural of the word rectus is recti.)

All of the seven muscles have their origin at the apex of the orbit around the **optic foramen** excepting the inferior oblique which has its origin on the nasal side of the floor at the base of the orbit in front. The four recti muscles extend forward from the apex an equal distance apart and are attached to the sclerotic from 6 to 8 mm. from the margin of the cornea. The superior oblique muscle which also has its origin at the apex extends forward close to the nasal side of the orbit to the internal angle at its base and then through the small tendinous pulley called the **trochlea**; from there is extends obliquely backward and over the upper and middle part of the eye where it becomes attached to the sclerotic underneath the superior rectus a little back of the equator. The inferior oblique from its origin under the nasal side of the front of the orbit, passes below the inferior rectus and turns up on the temporal side between the sclera and the external rectus and is attached to the sclera on its temporal side back of the equator of the eye.

The Intrinsic Muscles.

The muscles inside the eyeball are called **intrinsic** muscles and are·

1. The **ciliary muscle** or muscle of accommodation which surrounds the border of the lens.
2. The iris muscles, circular and radiating; the circular muscular fibres—**sphincter pupillae**—contract the pupil and the radiating fibres dilate the pupil—**dilator pupillae.**

The Nerve Supply

The principal nerves of the eye that are of interest to the Optometrist are the **third, fourth** and **sixth cranial nerves.** The third nerve, also called the **motor oculi,** supplies all the muscles of the orbit excepting two; the **superior oblique** which is controlled by the fourth nerve and the **external rectus** by the sixth nerve. Inside the eye the third nerve contracts the **ciliary muscle** and also contracts the pupil in the iris. The sympathetic nerve dilates the pupil. Practically all the contents of the orbit, the eyeball, optic nerve and muscles are enveloped in a fibrous sheath called the **capsule of tenon.**

Orbital fat
Inferior oblique m.
Lacrymal gland

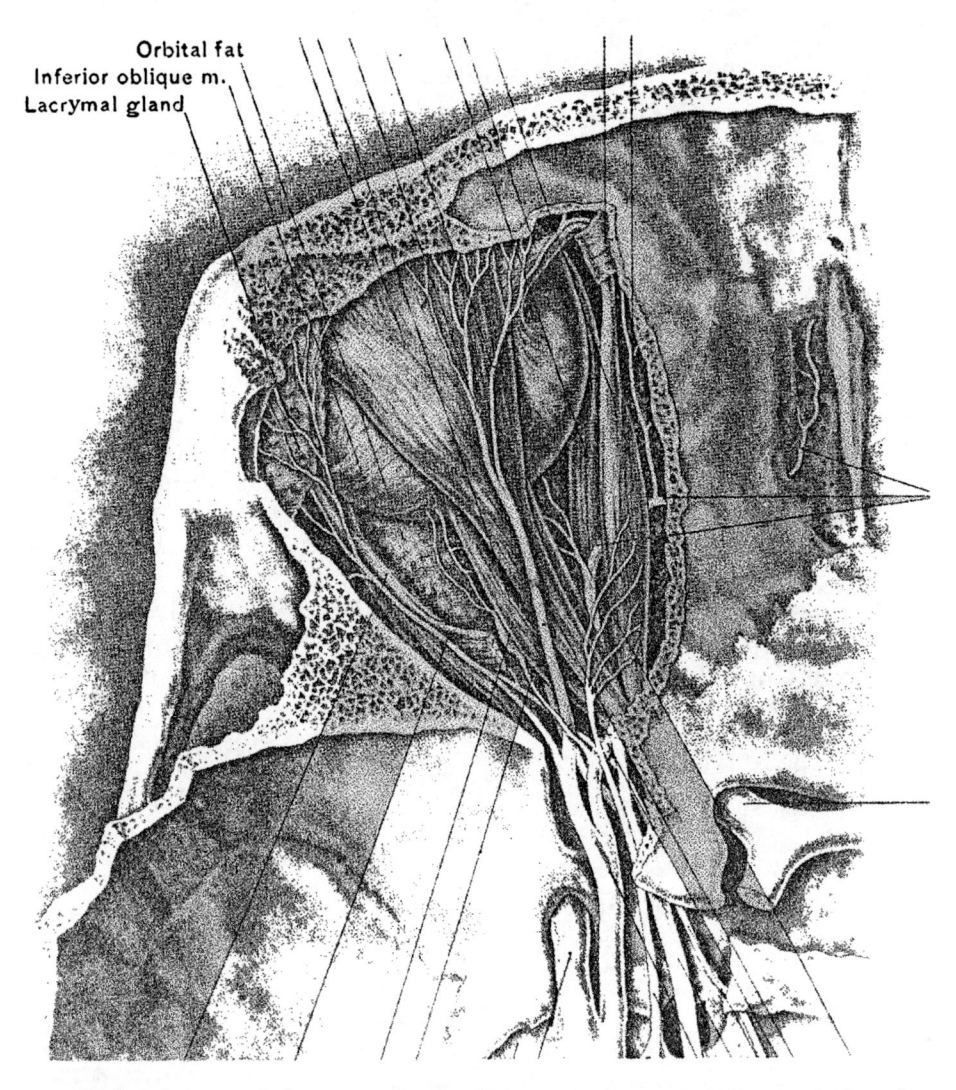

This picture—left eye—shows the general arrangement of the contents of the orbit as seen from above. Also the optic foramen and the optic nerve passing through it, as well as the sphenoidal fissure showing a direct open passage between the brain and orbital cavity, which serves as a passage for the cranial nerves, ophthalmic nerves, and small arteries.

The wide muscle at the top is the muscle that lifts the lid—levator palpebrarum.

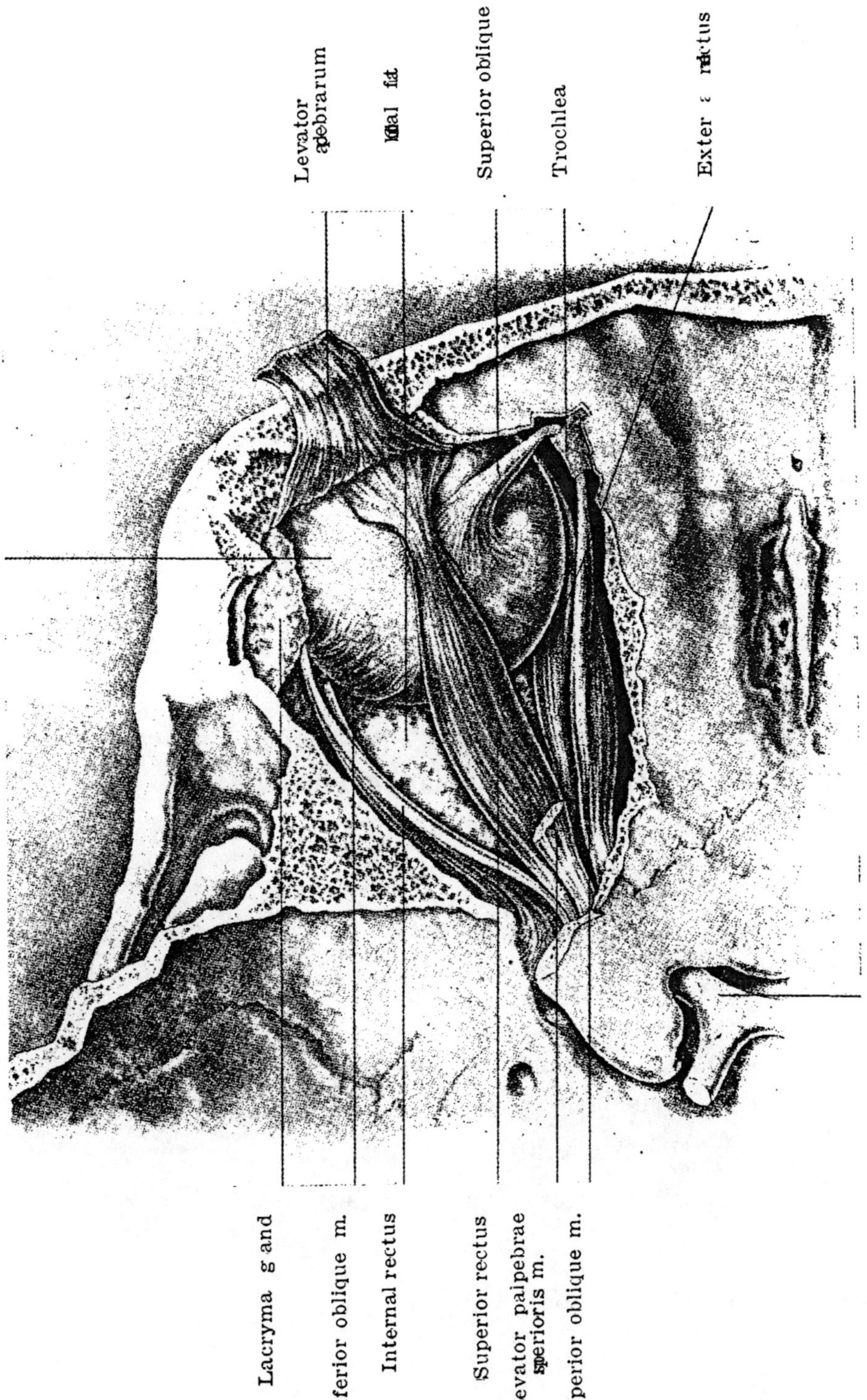
Levator
Superior oblique
Trochlea
Inferior oblique m.
Internal rectus
Superior rectus
Levator palpebrae
Superior oblique m.

Especial attention is drawn to the superior oblique muscle where it passes through the little tendinous ring called the trochlea, and from there turns at an angle backwards toward the eye-ball, where it widens out and becomes attached to the sclerotic a little back of the equator of the globe underneath the superior rectus.

It will be seen that the insertion of the inferior oblique muscle is on the temporal side and somewhat farther back from that of the superior oblique. These two muscles are classed as the rotary muscles whose principal function is to rotate or turn the eye on its axis. Note the shape of the superior rectus. Small at its origin, becomes wider, then narrow, and again wider in order to form a broad surface of at least 10 mm. where it is attached to the sclera in front of the equator about a quarter of an inch—6 to 7 mm.—from the cornea. This same description applies to the other three recti muscles.

QUIZ ON LESSON THREE.

1. How many muscles control the action of the eyeball? Name them.
2. What does the term "extrinsic muscles" mean?
3. Name the muscle that lifts the upper lid.
4. Give the origin and insertion of the two oblique muscles.
5. Give origin and insertion of the recti muscles.
6. Which side of the base of the orbit is the trochlea on?
7. Where do we find the intrinsic muscles? Name them.
8. Name the nerves that supply the motor power of the extrinsic muscles and which one of them also supplies the intrinsic muscles?
9. Study the two plates carefully.
10. See special anatomy on muscles in **PART TWO**.

LESSON FOUR. (Section One.)

THE EYELIDS—(Palpebrae).

The eyelids are two movable curtains placed in front of the eyeball to serve as a general protection from injury, dust, excessive light, etc. Along the margin of each lid are hairs called **cilia** or eyelashes. There are two or more rows of lashes in each lid, being longer and more numerous in the upper lid and curved somewhat upward, while those in the lower lid turn in the opposite direction, downward. These lashes should never be cut or trimmed as they serve to prevent small particles from getting in between the lids.

Each lid has the following arrangement of parts from the skin inward:

(1) Skin.
(2) Areolar tissue.
(3) Orbicularis palpebrarum muscle.
(4) Tarsal plate.
(5) Palpebral ligaments.
(6) Meibomian glands.
(7) Conjunctiva.

Palpebral fissure is the name given to the space between the edges of the open lids—(fissure means an elongated opening).

Conjunctival sac is the name given to the space that lies between the inside of the eyelids and the eyeball because of the membrane of that name (conjunctiva) which lines the lids and also covers the front part of the sclerotic. It is in this sac that small particles are held that get "into the eye".

Outer canthus is the name given where the edges of the upper and lower lids come together on the temporal side.

Inner canthus is where the lids join on the nasal side.

Canthi is plural of the word canthus.

Looking at the lids, closely, you will find at the inner canthus a little elevation or **point** on either lid where there is a small hole (see picture in **Lesson Five**). It is called the

Puncta larchrymalis.

This opening in each lid leads to a canal through which the ordinary supply of lachrymal fluid called tears, is drained off into the nose. There, also, at the inner canthus will be found a small fleshy spot with a few fine hairs in it. This is called the

Caruncle.

Close to the inner canthus and practically attached to the eyeball is a small fold of loose pink tissue which is called the

Plica Semi-lunaris.

(Plica means fold; semi-lunaris means half moon). It is also called the half moon fold. This membrane is the vestige of what in the early career of man was his third eyelid. It is still found fully developed in birds and some of the lower animals.

The different parts of the lids besides the skin may briefly be described as follows:

Areolar tissue means a tissue composed of white and yellowish fibres widely diffused throughout the body. Its function is to give strength and elasticity to a part as well as serve as a protection from injury. In the lids it lies next to the skin and acts as a sort of a cushion to protect the eyeball.

Orbiculus palpebrarum muscle is described in Section Two of this Lesson.

The **Tarsal plate** (tarsal cartilage) is a thin, cartilaginous tissue, which gives form to the lids, and, when both lids are closed, forms a shield for the eyeball in front. The cartilage of the upper lid is much larger than that of the lower, and at its upper margin is attached to the end of the muscle that lifts that lid, the levator palpebrarum.

The **Meibomian glands** are small sebaceous (fatty) glands imbedded in the substance of the tarsal cartilages and are placed side by side vertically in each lid from one canthus to the other. They number about thirty in the upper lid and a few less, say about twenty-five, in the lower. These glands have openings on the border of the lids along among the eyelashes. They secrete an oily substance which serves to lubricate the conjunctival sac.

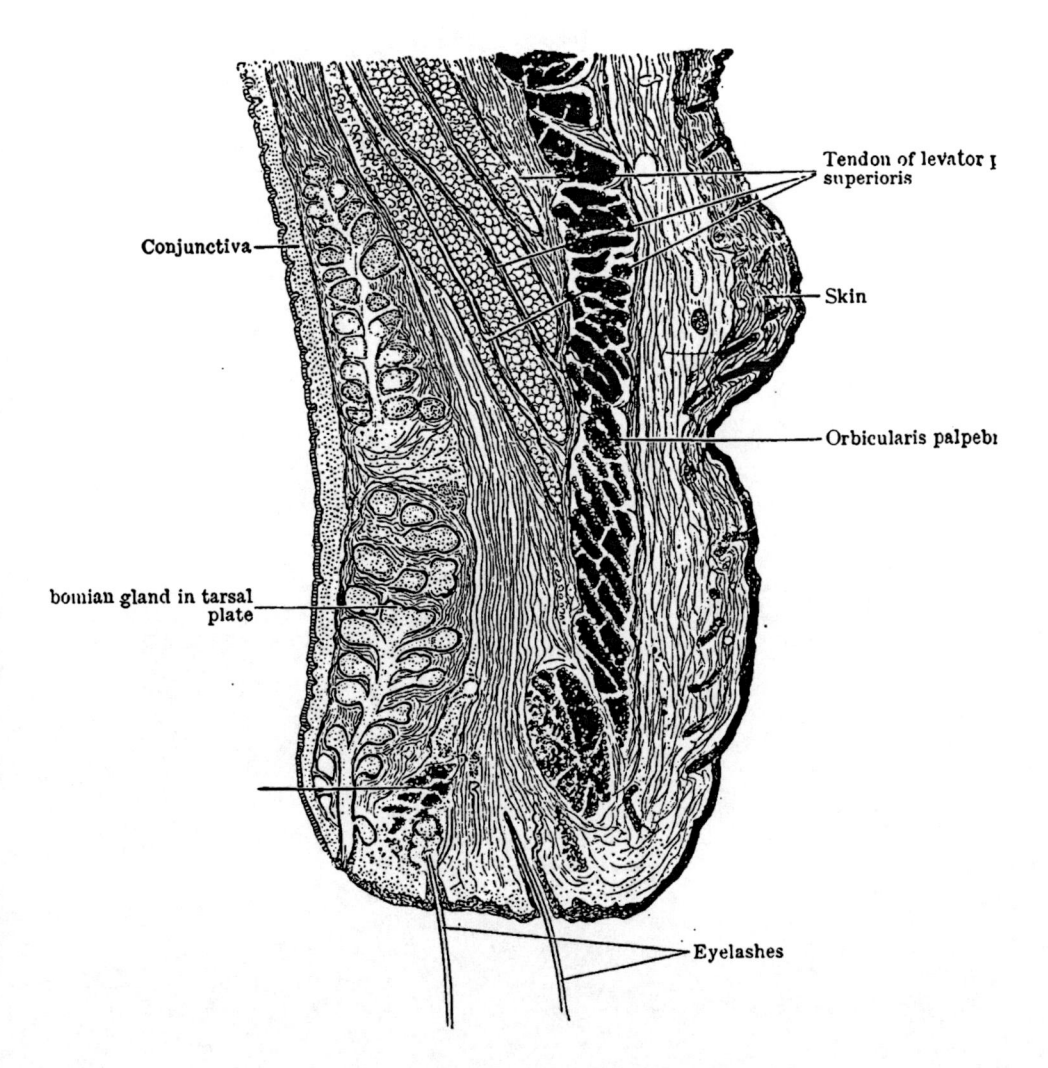

Showing the Structure of the Upper Lid.

The **Conjunctiva** is a thin mucous membrane which begins at the edges of the lids, lines them and folds back upon the sclera (to which it is loosely attached) and covers the front of the eye to the margin of the cornea. The part lining the lids is called the **palpebral conjunctiva;** where it folds back upon the eyeball it is called the **fornix** (arch); and that part which covers the sclera in front is the **ocular conjunctiva.**

LESSON FOUR. (Section Two.)

THE MUSCLES OF THE EYELIDS AND EYEBROWS.

Occipito-frontalis
Orbicularis palpebrarum
Tensor-tarsi
Tendo oculi
Corrugator supercilii

} All supplied by the seventh nerve, also called facial nerve.

Levator palpebra superioris (in **part**). Third nerve.

The **occipito-frontalis** is the forehead muscle. It elevates the eyebrows and produces wrinkling of the forehead.

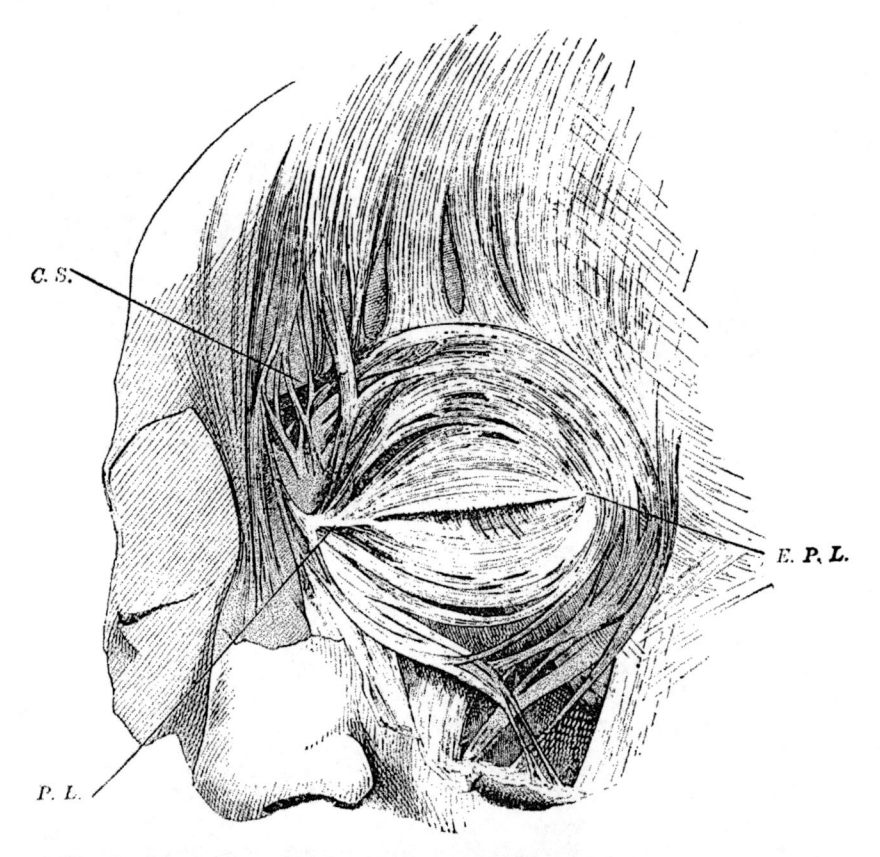

Orbicularis palpebrarum.—The palpebral and orbital portions are easily recognized, though the line of separation is not always to be seen. C. S. points to the corrugator supercilii; I. P. L., internal palpebral ligament; E. P. L., position of external palpebral ligament. (After Henle.)

The **orbicularis palpebrarum,** (sphincter oculi) is the chief muscle of the lids and is a powerful voluntary sphincter, consisting of an orbital, palpebral and lachrymal portion. It is a thin, flat muscle which lies immediately under the skin, encircling the eye and has fibres branching out connecting it with the brow, forehead and cheek. By its action the lids may be partially or gently closed or they may be tightly squeezed together.

The **tensor-tarsi** or **Horner's muscle** is a thin muscular sheet situated at the inner angle of the orbit **behind** the lachrymal sac. This muscle is really a deep portion of the orbicularis palpebrarum. It divides the two portions which cover the **posterior** part of each canaliculus. In **front** of the lachrymal sac is the **tendo oculi,** a short tendon about 6 mm. long and can be felt as a little ridge by pressing the finger against the side of the nose at the inner canthus. The tensor-tarsi and the tendo oculi both serve to empty the lachrymal sac by involuntary compression, thereby forcing its contents down through the nasal duct and from there into the nose.

The **corrugator supercilii** is a short ribbon-shaped muscle located at the upper ridge of the frontal bone at about the middle of the eyebrow. Its action is to draw the middle of the eyebrow inwards and downwards which gives the frowning aspect to the face.

The **levator palpebra superioris** acts in opposition to the orbicularis and elevates the upper lid. It has been mentioned in connection with the intra orbital muscles as it lies entirely within the orbit.

QUIZ ON LESSON FOUR.

1. What is the anatomical name for eyelid?
2. Locate the palpebral fissure. State what it refers to.
3. What is meant by outer canthus and inner canthus?
4. State just exactly what constitutes the conjunctival sac.
5. Make a diagram showing in the proper position, the puncta, caruncle and the half moon fold.
6. What does the plica semilunaris represent?
7. Which way do the eyelashes turn in each lid?
8. Name the layers of the lid from the skin inward.
9. Which layer gives form to the lids?
10. How many meibomian glands in each lid and what is their function? What separates them from the eyeball?
11. Describe the conjunctiva; the fornix.
12. Locate the palpebral conjunctiva; ocular conjunctiva.
13. Name the principal muscles of the lids and eyebrows.
14. Where do we find the orbicularis palpebrarum? What nerve contracts it?
15. Just where is the tensor-tarsi and tendo oculi relative to the lachrymal sac? What are they for?
16. Where is the corrugator supercilii and what is its action?

LESSON FIVE.

THE LACHRYMAL (Lacrimal) APPARATUS.

This refers to the secretion of "tears" and of their disposal, and is divided into a
Secretory portion and an
Excretory portion. The former consists principally of the lachrymal gland; but as a matter of fact, the moisture that commonly cleanses the ocular and conjunctival surface comes from the mucous follicles of the palpebral conjunctiva, while more copious supplies of tears are furnished by the lachrymal gland. Patients often complain of dryness of the eye, the lid seems to stick to the ball. This is the result of the conjunctiva being affected so that the normal secretion is somewhat lessened. There are
two lobes to the gland; an
upper lobe oval in shape, about 20 to 25 mm. in length and 12 to 14 mm. in thickness. (See cut showing apparatus of right eye). The
lower lobe is smaller and is sometimes called the
accessory gland. The gland is located at the base of the orbit at about the
superior external angle, lodged in a depression of the frontal bone to which it is attached by loose connective tissue; the under surface rests upon the eyeball at the
fornix (or fold of the conjunctiva where it turns from the upper lid back upon the eyeball).

A study of the pictue will show the general plan of the entire apparatus. From the gland will be seen several little tubes—about ten—which connect the gland with the surface of the eyeball through the thin tissue between them and it is through these
Lachrymal ducts that the slightly alkaline fluid—the tears—is sprayed onto the globe.

The **Excretory Part** is located at the
inner canthus and consists of the parts that drain the tears off into the nose. These parts are the
Puncta lachrymalis (lachrymal point)·
Canaliculus (canal) plural, canaliculi;
Lachrymal sac;
Nasal duct.

The **puncta lachrymalis** is a very small opening on the edge of the lid connecting with the
canal about 7 or 8 mm. in length which are directed toward the nose where the upper canal and the lower one meet and form a common canal that connects with the lachrymal sac. From here the drainage is downward through a connecting tube, the
nasal duct which is directly continuous with the sac and leads into the nose.

The tears are drawn into the sac from the inner canthus where they settle in a little depression, the
Lacus lachrymalis, by suction, the motor power being supported by the
tensor-tarsi or
Horner's muscle, which consists of two parts of about 12 mm. long arranged so as to compress each canal, which they do involuntarily, and very often, thus sucking the secretion through the puncta. In excessive secretion of tears, as in crying, there is, of course, an overflow upon the cheeks. This is called
lachrymation.

Superior portion of lacrymal gland

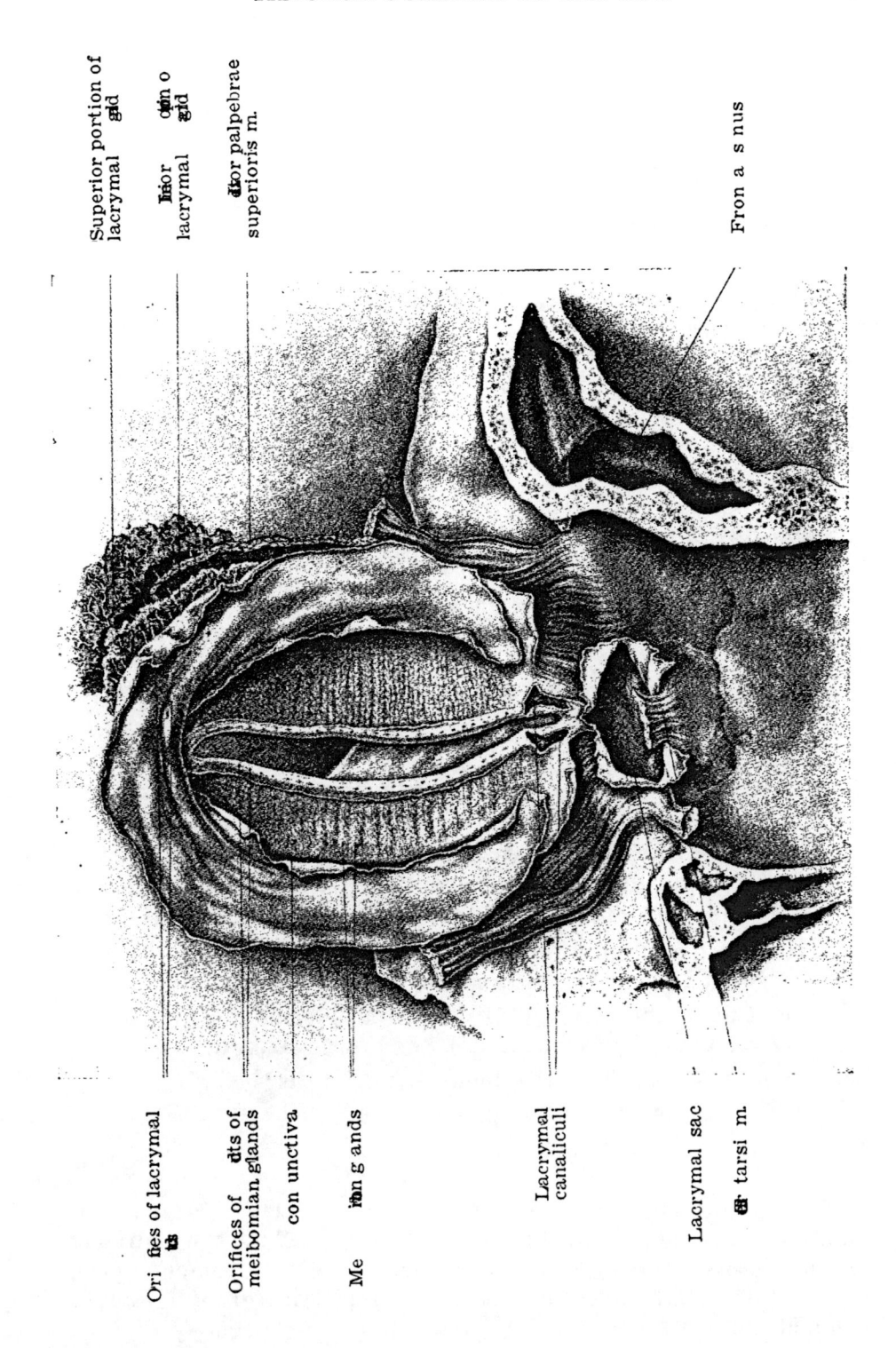

Inferior lacrymal gland

Levator palpebrae superioris m.

Frontal sinus

Orifices of lacrymal ducts

Orifices of ducts of meibomian glands

conjunctiva

Meibomian glands

Lacrymal canaliculi

Lacrymal sac

Orbicularis tarsi m.

The picture illustrates the meibomian glands as well as the lachrymal apparatus. Near the lachrymal gland will be seen 6 or 7 little spots which are to show where and how the tears get from the gland through the conjunctiva onto the eyeball. Just here the picture is somewhat misleading, as the conjunctiva is made to appear to be on the outside, or external to the meibomian glands instead of inside next to the eyeball.

The little dots along the edges of the lids are to represent the ordinarily invisible opening—meibomian follicles—at the ends of the meibomian glands through which an imperceptible oily secretion passes to lubricate the conjunctival sac, thus permitting the eye to move about without friction with the lids.

QUIZ ON LESSON FIVE.

1. What is meant by the term "lachrymal apparatus"? Give its divisions.
2. Why do people complain of dryness of the eye?
3. Describe the position of the two lobes of the lachrymal gland.
4. How and through what medium does the secretion from the gland get onto the eyeball?
5. Diagram and name each part of the excretory apparatus.
6. How is drainage affected? (See **Tensor Tarsi** and **Tendo Oculi**, lesson 4).
7. Do the meibomian glands bear any relationship to the lachrymal apparatus? If so, how? If not, why not?
8. Where do we find the lacus lachrymalis?
9. What is the plural for canaliculus?
10. Define lachrymation.

Note to the Student.—You have now learned a great deal about the eye just from these few lessons. Now take anyone of the pictures and talk to it and then show it to someone else, telling him what each part is and what it is for and how it connects up with some other part, etc.

PART TWO.

Part One covers the anatomy of the orbit, the eye, and its appendages in a general way so that the student can get a fairly complete conception as a whole of what is expected of him to master. However, there are some essentials that he should know, as well, that are not found in any text on the eye, and would require some research to cover. These will be found fairly well presented in **Part Two**.

PART TWO.

GROSS DESCRIPTIVE ANATOMY AND PHYSIOLOGY OF THE EYE.

THE SCLEROTIC.

This is sometimes called the Sclera or the white of the eye and is an opaque fibrous membrane covering five-sixths of the entire eyeball. In old age it sometimes becomes a dull yellowish hue, due to infiltration of fat, especially near the margin of the cornea. Its greatest thickness is found at the back part around the optic nerve where it is about one mm. From this region forward it grows thinner until it is from four-tenths to six-tenths of a mm. only. Where the sclera and cornea come together is called the **sclero-corneal margin.**

It is scantily supplied with blood-vessels and consequently gets its nourishment from its own lymphatic canals with which it is abundantly supplied. Its **nerves** are derived from the ciliary nerves.

THE CORNEA.

The cornea is regarded as the front window to the eye and as an object glass of the ocular camera it is one of the most important portions of the apparatus. Being necessarily placed at the front, and exposed whenever the eyelids are parted, it is more frequently injured than any other part of the eye. It comprises one-sixth of the external tunic or coat of the eye and its essential features are as follows·

It has **five layers:**

(1) **Epithelium.**
(2) **Bowman's membrane,** or Anterior elastic lamina.
(3) **True corneal tissue** (Cornea Propria) which is the center layer and much thicker than any of the others.
(4) **Descemets membrane** or Posterior elastic lamina.
(5) **Endothelium.**

Diameter—Vertical 11 mm.; horizontal 12 mm.

Refractive power—about 42 dioptres.

Radius of curvature—7.8 mm. horizontal meridian and 7.7 in the vertical.

Index of refraction—1.33.

Blood vessels—none.

Nerves—highly sensitive—60 to 80 branches of the anterior ciliary nerves enter the cornea.

The cornea attains its permanent dimensions very early in life and varies but little after the third year. It developes faster than the rest of the eye.

Nutrition—As the cornea has no blood vessels from which to get nutrition and grossly speaking is a network of cells, it maintains itself upon the vital force of these cells, causing an inter cellular flow of lymph which remains about equal during life.

Some interesting experiments have been made to determine the behavior of the cornea with regard to the rays of the invisible portions of the spectrum. Its power of absorption of the infra-red or heat rays is a little superior to that of water, but not notably so. The chemical or ultra-violet rays also appear to pass through the cornea without sensible diminution.

ARCUS SENILIS.

In elderly persons there is often seen a narrow gray cresentric line either around the cornea or at its upper border. This is called arcus senilis, or the arch of old age. It never interferes with the vision, although it may extend some distance toward the center. It is occasionally seen in young people, but is usually not seen before fifty or sixty years of age, owing to decrease in nutrition with advancing years.

THE IRIS.

The iris is a colored membrane, circular in form, hanging behind the cornea directly in front of the lens and in contact with it and perforated at about its center by an aperture of variable size called the **pupil.**

In new-born white children the iris is almost always blue. This is due to the fact that its pigment-cells do not develop until sometime after birth, the coloration not being complete until after the second year. In Albinos the pigment is entirely absent. The distribution of pigment varies greatly in different individuals.

Diameter—10 to 12 mm.

Thickness—.4 mm.

Diameter of pupil ranges from 3 to 6 mm.

Blood supply—The blood vessels of the iris come from the two branches of the Ophthalmic artery, known as the long posterior ciliary arteries, also the anterior ciliary arteries.

Nerve supply—The contraction of the pupil occurs by the action of a branch of the third nerve upon a narrow muscular band, called the
Sphincter pupillae which encircles its border; and dilation occurs by relaxation of the sphincter and contraction of the radiating muscular fibres called the
Dilator pupillae, which action is controlled by the sympathetic nerve.

The structure of the iris presents two chief layers—the iridial stroma or body proper and the pigment layer: these include five sub-layers—

1. Anterior endothelium.
2. Anterior boundary layer.
3. Vascular stroma layer.
4. Posterior limiting layer.
5. Pigment layer.

The vascular stroma layer, forming the bulk of the iris, consists of loose connective tissue supporting the numerous blood vessels and nerves which occupy this layer.

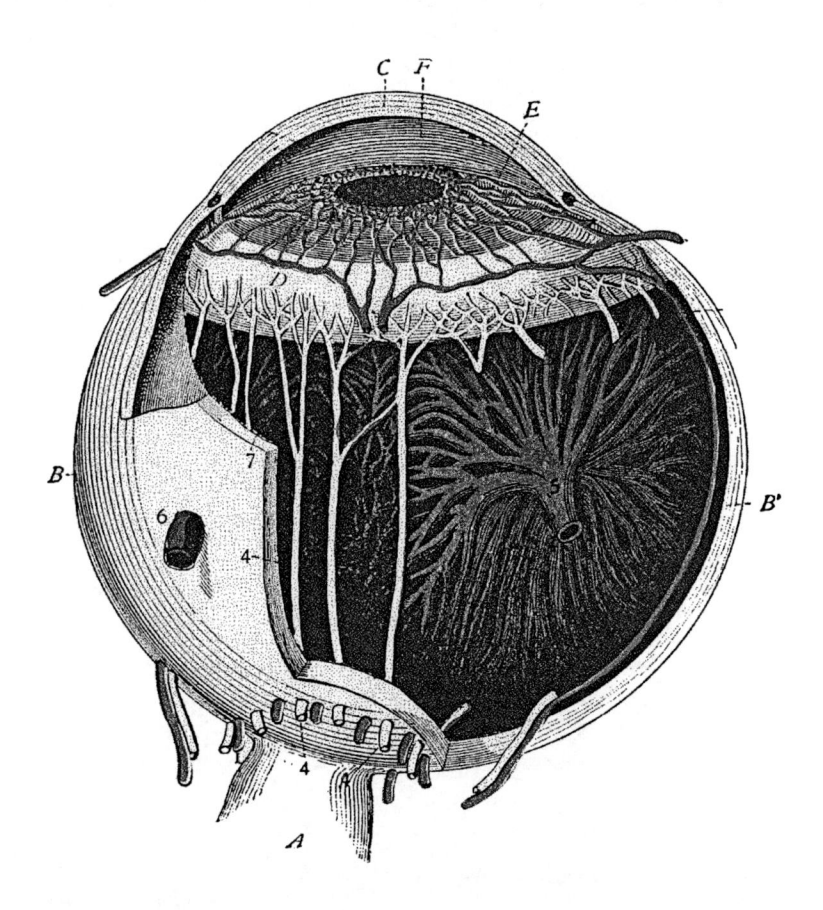

This picture is not of the eye itself, but is a diagram made to show why the second coat is called the **vascular coat.** The word vascular means **tube or tubes.** Blood vessels and nerves are tubes. The general color of the choroid is brown, not blue. However, it is customary in coloring anatomical pictures to show the arteries red because the blood within them **is bright red**, while in the veins it is a much darker red, and as seen through the skin presents a bluish tinge. For general blood supply see "Ophthalmic Artery".

THE CILIARY BODY.

The ciliary body is that portion of the second tunic directly back of the iris and extending back to the choroid. It consists of two parts—**ciliary muscle** and **ciliary processes**, which form a sort of a ring around the margin of the lens. The ciliary muscle being close to the sclera near the sclero corneal junction, while the processes are a little farther back or under. It is supplied by a branch of the third nerve and possesses the involuntary function of adjusting the convexity of the lens—called accommodation. The ciliary muscle contracts and pulls the ciliary processes forward toward the lens, thus relaxing the tension on the suspensory ligament which holds the lens; with the tension relaxed the pressure is removed off the anterior surface of the lens which then assumes a more convex condition, sufficient to keep images upon the retina at different distances at which the eye may be directed.

The ciliary processes are some seventy or eighty slight irregular folds and are reallly the forward continuation of the choroid and it is to these that the suspensory ligament is attached. They are the most vascular portion of the eyeball, principally composed of pigment and numerous blood vessels and this body is the principal source of the aqueous humor.

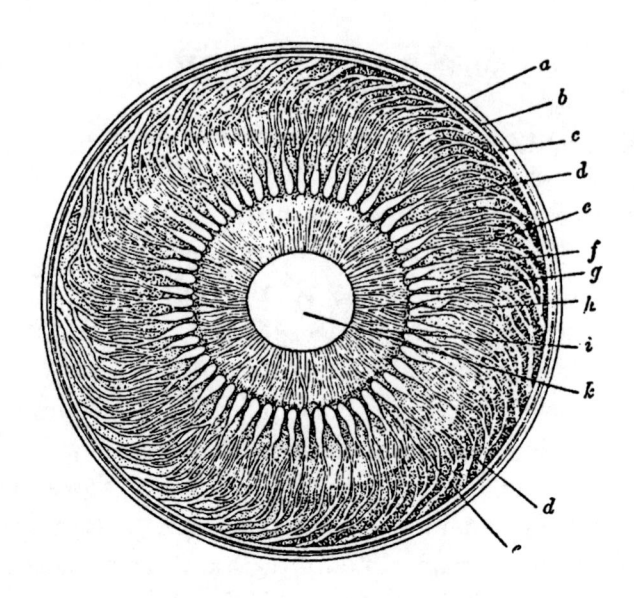

Study this diagram carefully as it represents a cross section of an eye cut through just back of the ciliary body and shows the relation of one part to any other part of that region of the second coat as indicated by lines. The student will disregard a. b. c. and d. and begin at the white spot in the center, the pupil. From the margin of the pupil to the next ring is the iris. From this ring—the iris—the ciliary body begins and extends backward toward the choroid about 6 mm. Close to the iris you see what looks like a round string of elongated beads to illustrate the ciliary processes. The letter g. points to the corona radiata, which means the iris sets in a raised ring, and it is in this ring that the lens belongs. In the study of the lens and in accommodation this picture should be kept in mind.

THE CHOROID.

The choroid is a dark brown membrane lying between the sclera and the retina and constitutes the posterior two-thirds of this second coat or tunic from the ciliary body back. It is very thin, varying from .06 of a mm. in front to about .1 of a mm. at the back. It is also called the **vascular coat** because it consists mainly of blood vessels which are united by delicate connective tissue containing numerous pigmented cells. The arteries are the short ciliary. Its function through its vessels is chiefly to serve nutrition to the retina, vitreous and lens. It forms the dark coating of the interior of the eyeball and its dark pigment is nature's provision to modify the intensity of light that enters through the pupil.

THE RETINA.

The retina is a very thin delicate membrane which consists principally of an **expansion of the optic nerve.** It is the inner coat or tunic and extends forward to the ciliary body where its termination is called the **ora serrata.** From there on, devoid of nerve fibres and much thinner, it is continued on forward over the inner surface of the ciliary body and posterior surface of the iris. In the living eye it is transparent and of a purple red color; after death it soon becomes opaque. It is connected with the choroid at the entrance of the optic nerve at the back and at the ora serrata in front, otherwise it simply lies upon it, but is not attached to it. On this account vision is often destroyed by detachment of the retina from its position against the choroid.

The minute anatomy of the retina is very complicated. It is the complete development of this part of the eye that is especially necessary to good vision. Vibrations of light reach it from all directions in front of the eye, but its region of most distinct vision is about 1 mm. to the temporal side of the optic axis. This is called the **macula lutea** or **yellow spot** which is slightly oval and approximately 2 mm. in its great diameter which is horizontal. Near the center of the yellow spot occurs a small depression known as **fovea centralis** or center of focus. From the fovea to the center of the optic disc it is about 4 mm., the optic disc being about 1.5 mm. in diameter.

The retina has ten layers in the order named from the choroid inward.

1. Pigmentary layer.
2. Layer of rods and cones.
3. External limiting membrane.
4. External nuclear layer.
5. External granular layer.
6. Internal nuclear layer.
7. Internal granular layer.
8. Vascular layer.
9. Fibrous layer.
10. Internal limiting membrane.

The second layer is of the most interest because upon its proper development depends the best visual acuity. At birth there are 3,360,000 cones and about 130,000,000 rods in this layer and it is upon the further development of the eye that good vision depends. Should anything interfere with the complete growth of the retina to prevent development of the number of cones to the extent of about 7,000,000 the vision is never perfect and cannot be made so with glasses. In the distribution of these cones it has been found by microscopical examination that from the ora serrata back toward the macula they gradually become more numerous and closer together until within the macula there are about 13,000 cones and no rods at all.

The rods and cones are the terminal organs of the optic nerve; receive vibrations of light which fall upon the retina and connect these virbrations into impulses which are carried by the different branches of the optic nerves and tracts to the brain; here they produce the sensation of light. When an image falls upon any other part of the retina there is indistinct vision.

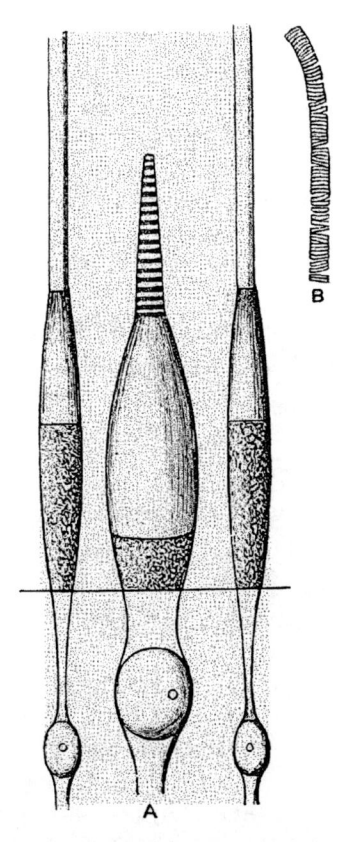

A, A cone and two rods from the human retina (modified from Max Schultze); B, Outer part of rod separated into discs.

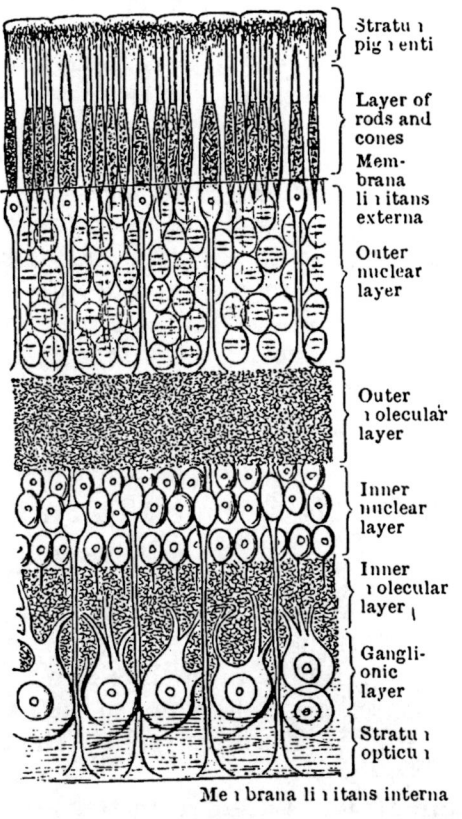

Diagramatic Section of the Human Retina (modified from Schultze).

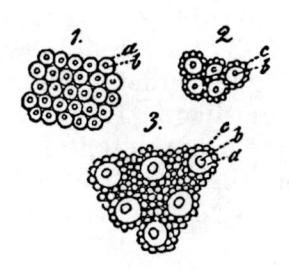

Stratum pigmenti
Layer of rods and cones
Membrana limitans externa
Outer nuclear layer
Outer molecular layer
Inner nuclear layer
Inner molecular layer
Ganglionic layer
Stratum opticum
Membrana limitans interna

Surface view of retina, showing disposition and relative number of the rods and cones. (Kolliker.) 1, from the fovea—only cones; 2, from the margin of the macula lutea; 3, from midway between the fovea and the ora serrata; a, profile of larger inner segment; b, of smaller outer segment; c, rod.

THE OPTIC NERVE.

The optic nerve is regarded as part of the brain, and is devisable into three portions, **cranial, orbital** and **ocular** portions. It is about 50 mm. long from the eye to the optic commissure (also called the optic chiasm), where it meets the optic nerve coming from the other eye. It is 30 mm. in the orbit, 10 mm. in the optic canal at the apex of the orbit and 10 mm. intracranial (within the cranium or skull). Behind the commissure the two optic nerves become the optic tract. The nerve is about 5 mm. across. It has the form of a modified S as it lies in the orbit, thus allowing the eyeball to move about without tension on the nerve.

OPTIC DISC.

Optic Disc. Optic Nerve entrance. Optic papilla. Nerve head. Blind spot. (Sometimes improperly called Porus Opticus.) This is the termination of the optic nerve as it pierces the eyeball and spreads out to form the inner layer of the retina—the internal limiting membrane. Normally, the optic disc is nearly circular in outline and is about 1.5 mm. in diameter. It is located about 3.5 mm. to the nasal side and about 1 mm. above the line of fovea centralis. The optic axis of the eye being between these two points. It has a pinkish tint and on careful examination, is seen to present differently colored zones. (1) A central clear spot, which is the funnel-like depression from which emerges the central retinal vessel. (2) A vascular zone containing many capillaries. (3) A narrow light band, which is the connective tissue ring. (4) Surrounding all, the dark choroidal zone. Its size as seen with the ophthalmoscope, direct method, in the emmetropic eye is said to be fifteen times larger than actual. In the hyperopic eye it is comparatively smaller and in myopia greater. Whoever figured it out that the magnification was fifteen times must have overlooked the actual facts in the case. If the disc is 1.5 mm. in size and we multiply that with 15 we get a diameter of 22.5 mm. or almost an inch. Never yet have I seen a disc look anywhere near that size. At most it appears to be about 17 mm.

PORUS OPTICUS.

Porus opticus is the physiological excavation and passage on the nasal side of the optic disc where the retinal vessels are seen.

CAPSULE OF TENON

(Note—See "orbital fat")

Carefully study this, as, usually given in text books, it is somewhat difficult to understand just what the capsule really means. Look at the pictures shown here as you read this. The Capsule of Tenon, also called **oculo-orbital fascia,** insheathes all the organs which pass through it and forms a **cup for the eyeball, is continuous with the sheath of the optic nerve** and also **forms a secondary attachment for the ocular muscles.** It is a delicate opaque membrane. While it appears as a part of the eyeball, it is not, as there is a lymph space between it and the sclerotic which it covers to within 3 mm. of the sclero-corneal margin where it fuses with the ocular conjunctiva which covers it. It is taught by common consent that Tenon's Capsule is a socket in which the eyeball rotates without change of position, (meaning the socket is immovable and the ball moves about in it when the extrinsic muscles pull in it in any direction). Anatomy shows that this is out of the question because the anterior part of the capsule is closely attached to the sclera in front of the insertion of the recti muscles close to the cornea, hence the **two move together** upon the cushion of fat behind them. In looking at the picture it will be seen that the fibrous tissue of the sheaths of the muscles is continuous with that of the socket, the effect of which is partly to steady the eyeball and to resist the backward pull of the muscles. It will be understood that when upon operating for strabismus a muscle is cut entirely free from its insertion, it cannot drop out of position, but retains its relationship with the other extrinsic muscles.

The **Check Ligaments** aid in this also.

CHECK LIGAMENTS

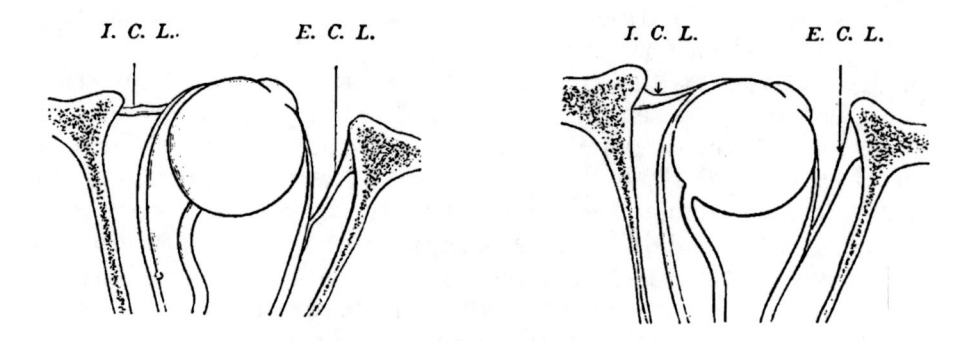

I. C. L.   *E. C. L.*   *I. C. L.*   *E. C. L.*

The check ligaments during **partial** contraction of the external rectus muscle, the internal check ligament (I. C. L.) being in a state of maximum relaxation, and the external (E. C. L.) somewhat stretched. (Motais.)

Diagram intended to show how, during full contraction of the external rectus, the external check ligament (E. C. L.) is stretched to its maximum length, and the internal (I. C. L.) is slightly stretched also. (Motais.)

Note.—It will be observed that as the eye is shown turned to its utmost there is no pull on the back part of the ball by the optic nerve because of the fact that the nerve is very flexible and of a modified S shape, which admits free movement. Most pictures showing the optic nerve make it appear to be practically straight, which is not true.

The depth of the orbit is 45 mm., the eye-ball 24 mm., sets in it leaving 3 or 4 mm. at the base, thus from the back of the ball to the optic foramen is only a little over 17 mm., and as there is 30 mm. of the optic nerve in the orbital cavity from the eye-ball to the foramen, that leaves over 10 mm. for rotation.

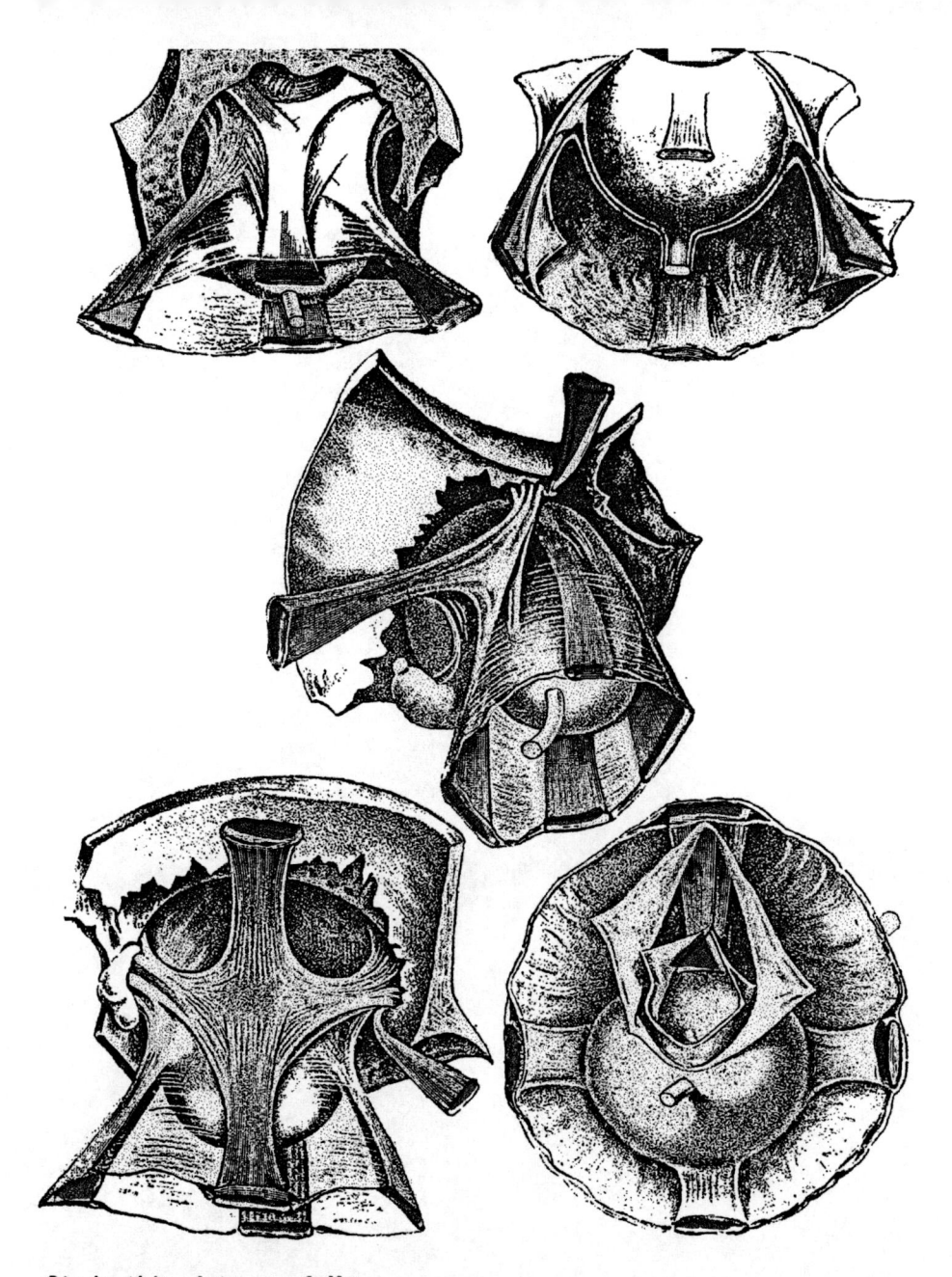

Study this plate carefully.

It shows how the entire eye-ball and the muscles as well as the spaces between one muscle and another is enveloped in the same membrane. At the ends of the muscles, where cut, it will be seen that each one is completely enveloped; also how the eye-ball "sets in" the so-called cup and that a little to the anterior of the middle of the ball the membrane turns back on the inside of the muscle and continues to envelope it (them) to its origin at the apex of the orbit. The lower left Figure presents another aspect of the enveloping membrane. In the lower right Figure another view directly from behind the eye, forward, emphasizes the fact of the complete enveloping membrane called **The Capsule of Tenon.**

AQUEOUS HUMOR.

This is a thin clear alkaline fluid occupying the anterior and posterior chambers, and is supplied by the ciliary processes. Its index of refraction is 1.33. In case of an injury or operation resulting in loss of aqueous the cavity refills in a few moments. It does not seem however, that under ordinary natural conditions a very rapid secretion of the aqueous takes place because its principal source of exit is through the spongy tissue of the **spaces** of **Fontana** at the sclera-corneal margin where it is drained off through the **canal** of **Schlemm** by the anterior ciliary veins. To a lesser extent it also passes out by the lymph-crypts of the iris.

The extraordinary solvent properties of the aqueous humor makes it easily affected by drugs circulating in the blood. Should the lens substance come in contact with it in small portions at a time it completely dissolves it. It is by this method that soft cataract in children is treated and the lens substance is made to gradually disappear.

THE VITREOUS BODY OR HUMOR.

is a soft gelatinous, perfectly transparent substance and occupies the posterior cavity called **the vitreous chamber** also **hyaloid cavity.** It has no special value in refraction excepting its **index** of **refraction** which is 1.33. It is contained in a very thin transparent capsule—the **hyaloid membrane**—which separates it from contact with the retina. It gets its nutrition from the choroidal vessels. A certain amount of vitreous may be removed without seriously injuring vision, and it seems to be rapidly renewed from the ciliary processes through the **zonula** of **zinn.** The only well established exit of fluids from the eye—the aqueous and vitreous—is that at the angle of the anterior chamber.

THE LENS AND ITS CAPSULE.

Before birth while gradual development of the eye is taking place, the lens is supplied with its nourishment by a vascular membrane which surrounds and covers it. The vascular portion of this membrane gradually disappears as the lens completes its development leaving it entirely clear and it thereafter serves as a capsule or complete cover for the lens and protects it from the surrounding aqueous humor.

The lens is held in position by the **suspensory ligament,** also called the **zone** of **zinn** and **zonula of zinn,** which is the thickened portion of the hyaloid membrane extending from the ciliary body to the margin of the lens on its anterior surface.

This is the membrane that is affected by the ciliary muscle when "accommodation" takes place. The lens is a biconvex circular body, lying directly behind the iris and in contact with it. The center of the anterior surface of the lens is its anterior pole, and is about 2.3 mm. from the back of the cornea and the center of the posterior surface is its posterior pole which is about 15.6 mm. from the retina. It is a little greater in convexity behind than in front. The central portion of its anterior surface is opposite the **pupil.** Its posterior convex surface lies against the hyaloid membrane forming a depression called the **patellar fossa** or **hyaloid fossa.** It is soft, elastic and transparent and is about 8.5 mm. in its transverse diameter and about 3.4 mm. thick at its least convexity and 4, at its greatest.

Radius of curvature. (Anterior) (Posterior)

At distant vision, least curvature............... 10 mm.   6 mm.

At closest vision greatest curvature......... 6 mm.   5.5 mm.

Refractive power about 16 D. Index of refraction 143.

Nutrition of the lens is supplied from the ciliary body.

ENLARGED DIAGRAM OF THE LENS

No. 1.

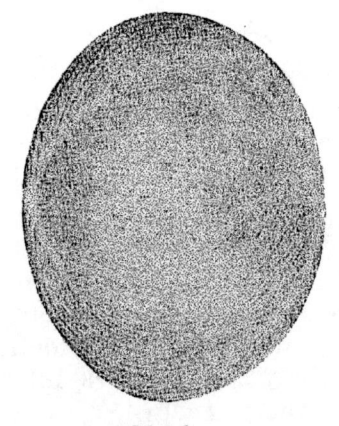

No. 2   No. 3

No. 1 shows the sectional layers of the lens which is somewhat similar to that of an onion, and opening up in its antero-posterior diameter.

No. 2 shows the relative proportions and curviture of both surfaces of the lens in its antero-posterioor diameter which is from 4 to 4.5 mm.

No. 3 shows the greatest diameter of the lens (about 8.5 mm.) as it is held in position directly back of the pupil.

The usual diagrams of the lens seen in books show it to appear oblong and sharp on the edge. It must be remembered that such a picture is made to represent the eyeball—a sphere—cut in half, thus leaving a flat side view. The lens cut in two, vertically, would appear oblong accordingly. Its edge is rounded—not sharp—and its surfaces always spherical, as shown in No. 2.

THE CANALS OF THE EYE

Petit's Canal, Hyaloid Canal, Schlemm's Canal

The **canal of Petit** is a narrow channel which encircles the margin of the lens. It is filled with lymph (a fluid) which comes from the ciliary vessels and is supposed to supply nutrition to the lens.

The **hyaloid canal**, also called the **canal of Stilling, canal of Cloquet** and **Central canal,** is a very fine line of space in the vitreous humor extending from the lens backward to the retina. It cannot be seen when looking into the interior of the eye with the ophthalmoscope.

The **canal of Schlemm** is located in the sclerotic close to the margin of the cornea forming a sort of a ring around the front part of the sclerotic. It is really a channel of small blood vessels which serve to carry off the debris of the eye back into the circulation. Directly where the iris and the cornea come together around the margin are a number of little openings called the **spaces of Fontana** through which the fluid passes from the anterior chamber in order to get into the **canal of Schlemm.** Whenever from disease or injury to the eye this canal is closed, the drainage of the eye is practically destroyed and the person gradually becomes blind.

ORBITAL FAT.

The orbit is filled with fat—adipose tissue—which is bounded in front by the capsule of Tenon and its fibrous expansions. It is very delicate in structure and forms an almost fluid support for the eye, well adapted for its movements in all directions without pressure. In operating for removal of the entire eyeball this fat is not disturbed, as the cutting is first made directly around the margin of the cornea where the conjunctiva and capsule are both dissected clear from the sclerotic and continued on to the insertion of the recti muscles when each one is raised with a hook and cut close to the sclera. The blunt pointed curved scissors continue to follow close to the sclera separating all tissue until the optic nerve is reached and cut, when the entire eyeball is then removed from the pocket or inside of the capsule. The muscles, the fascia and the fat have not directly been disturbed and of course retain their usual relationship and together form a basis for the use in wearing an artificial eye.

CIRCLET OF ZINN—LIGAMENT OF ZINN—ZONE OR ZONULA OF ZINN—TENDON OF ZINN

The **circlet of zinn** is the vascular circle around the optic nerve formed from twigs of the short posterior ciliary arteries.

The **ligament of zinn** is the lower part of the common tendon that encircles the optic foramen at the origin of the recti muscles and must not be confounded with the **zone of zinn** or **zonula of zinn,** which are other names for the suspensory ligament around the lens.

LIGAMENTUM PECTINATUM IRIDIS.

The **ligamentum pectinatum iridis** consists of a mass of spongy tissue and occupies the angle of the anterior chamber where it unites the iris and the ciliary muscle at the inner corneal border. It is intimately connected with the spaces of Fontana.

BLOOD SUPPLY.

The ophthalmic artery has been mentioned as a branch of the **internal carotid artery.** These two pictures will serve to show the principal blood supply of the head. It is called **The Carotid System of Arteries.**

These arteries are found on either side of the neck on about a vertical line with the ear. There is a main stem called the **Common Carotid.**

At a point just back of the lower jaw bone it separates into two branches which are named
The External Carotid and
The Internal Carotid.

Each of these again form several branches which have names according to the local parts they supply. The external is distributed about the external part of the neck and head while the internal is confined almost entirely to the contents of the cranial cavity. One other blood supply of the brain comes from the **Vertebral arteries.** It will be observed in the picture that the ophthalmic artery branches off from the carotid close to the apex of the orbit just back of the optic foramen and from there passes through the foramen along with the optic nerve into the orbit where it continues forward under the lower border of the superior oblique and its pulley—trochlea—to the base of the orbit where it terminates in two branches. Altogether the ophthalmic artery—the trunk—loses itself into ten separate branches and thus serves to supply individual parts of the contents of the orbital cavity.

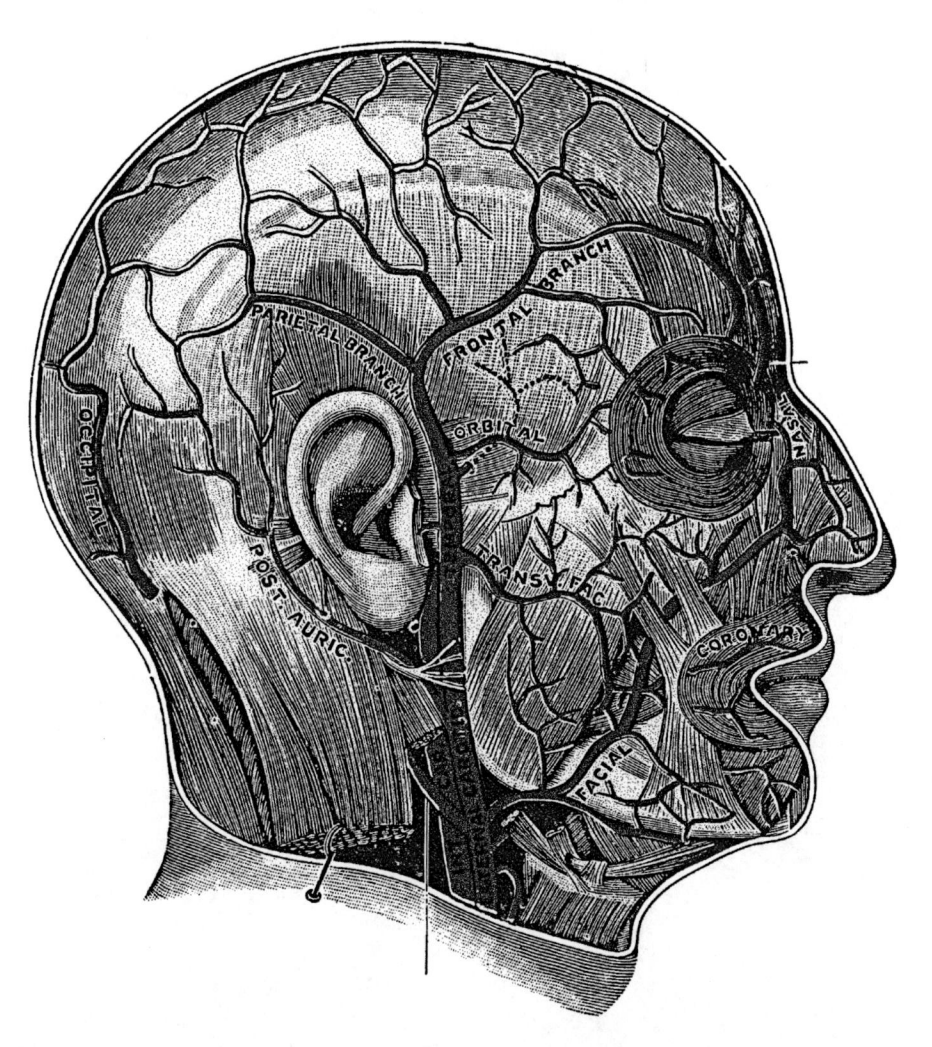

Showing location of the Cartoid arteries and where the Internal branches off and passes back underneath the ear and up into the skull.

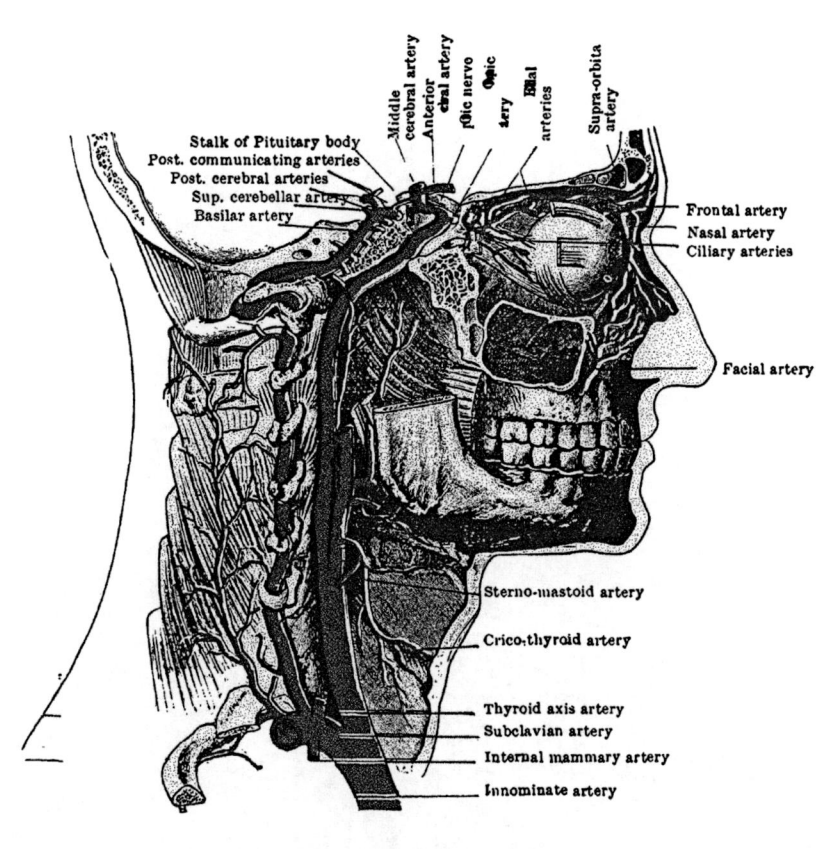

This picture is to show the connection of the Ophthalmic artery with the Internal Cartoid artery.

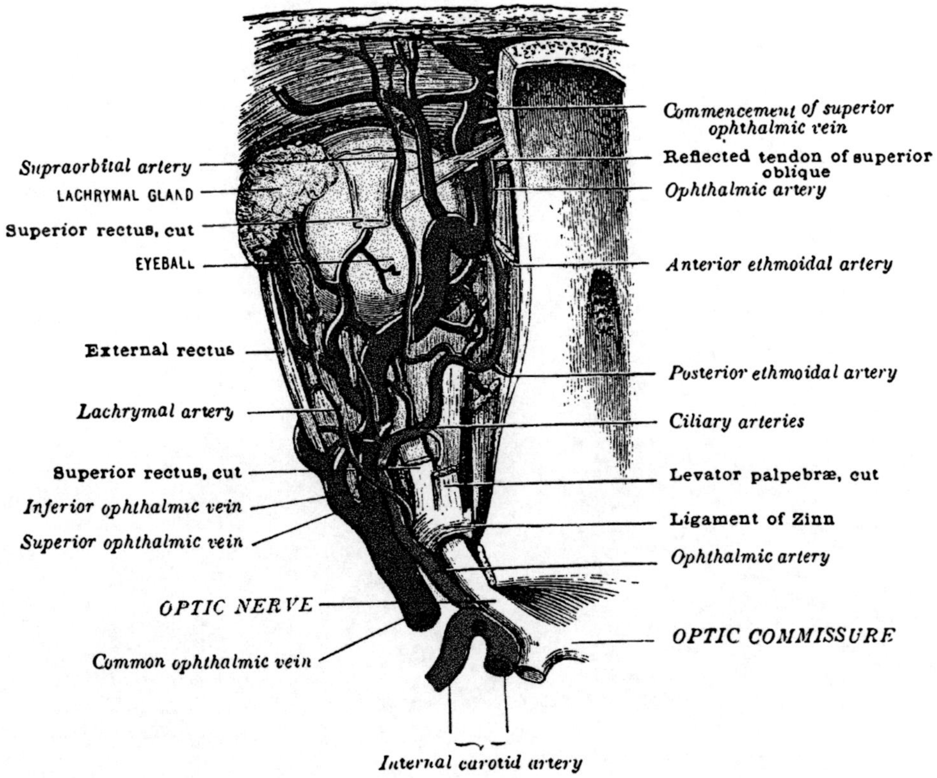
Supraorbital artery
LACHRYMAL GLAND
Superior rectus, cut
EYEBALL
External rectus
Lachrymal artery
Superior rectus, cut
Inferior ophthalmic vein
Superior ophthalmic vein
OPTIC NERVE
Common ophthalmic vein
Commencement of superior ophthalmic vein
Reflected tendon of superior oblique
Ophthalmic artery
Anterior ethmoidal artery
Posterior ethmoidal artery
Ciliary arteries
Levator palpebræ, cut
Ligament of Zinn
Ophthalmic artery
OPTIC COMMISSURE
Internal carotid artery

THE VEINS OF THE ORBIT

It will be remembered that the ophthalmic artery carries the blood **into** the orbit from the brain through the optic foramen. (See red vessels in the picture.) At the base of the orbit (in front) it finds its way into the veins (see **blue vessels** in cut) which gradually enlarge as they go back toward the apex until they form two main trunks—the **superior ophthalmic vein** and **inferior ophthalmic vein** which together at the apex form one single and larger vein—the **common ophthalmic vein** and from here passes into an opening called the cavernous sinus.

The branches of these veins are

(1) The superior muscular branches.

(2) The ciliary veins.

(3) The anterior and posterior ethmoidal veins.

(4) The lachrymal vein.

(5) The central vein of the retina.

THE EXTRINSIC MUSCLES.

Besides the ordinary text on these muscles there are several points that a refractionist should know. For extended reading involving every progressive thought of today on the action of these muscles the reader is referred to the two large volumes on "The Muscles of the Eye" by Lucien B. Howe, M. D., of Buffalo; also a single volume, "Motor Apparatus of the Eye," by George T. Stevens, M. D., of New York. Other books of minor value are of course on the market.

Not usually mentioned in connection with the extrinsic muscles there are **check ligaments** (ligamentous ailerous—orbital tendons) that should receive attention.

These names are given to small fibrous bands that connect each extrinsic muscle close to its insertion on the globe, to surrounding parts. They serve to modify any extreme action of the muscle proper, acting as bands of restraint as well as aid in harmonious action of two or more of the muscles and are an aid to perfect binocular fixation. See "Capsule of Tenon." In regard to the exact distance from the sclero-corneal margin the four recti muscles have their respective insertions, there is some little difference of measurements given by the authors because nature varies, but a fair average in detail is as follows:

	Length	Width at Insertion on Schlera	Distance from Cornea	Relative Power
Internal Rectus	41 mm.	10.75 mm.	6 mm.	Strongest
External Rectus	40.6 mm.	10.75 mm.	7 mm.	2d strongest
Inferior Rectus	40 mm.	13 mm.	7 mm.	3d strongest
Superior Rectus	41.8 mm.	13 mm.	8 mm.	Weakest

Your attention is now drawn to the **WIDTH** of the insertion of the muscles or rather the tendinous portion, in front, of each recti muscle of from 10 to 13 mm. Now added together the total distance around the eye-ball covered by the insertion of these four muscles is about 46 mm. As the eyeball at its equator is about 23 mm. that gives its greatest circumference say 70 mm. The antero-posterior diameter being about 24 mm., we now find that the insertion of these muscles being in front of the equator are at a point where the circumference is somewhat less—say 62 mm. This leads you to the fact that the combined length of these insertions practically make a complete band around the eye leaving only about 4 mm. between the margin of each insertion. From the insertion backward these muscles diminish in width, swell again at the center and become smaller again at their origin at the apex of the orbit.

The reason we do not see these muscles at their insertion is that they are covered first with the opaque capsule of Tenon which covers the sclera to within 3 mm. of the sclero-corneal margin and over this on the outside is the ocular conjunctiva.

Whenever an operation for strabismus is necessary the surgeon must first make an incision through both the conjunctiva and the capsule before he can get to the muscle.

Most of the pictures shown in the books are rather misleading in making the width of the insertion of the recti muscle appear rather narrow. The proper physiological function of the extrinsic muscles is to maintain fusion and therefore stereoscopic vision at any and all distances.

Action.—No one of the muscles of the individual eye acts singly, but in groups of two or more as shown in the following table:

Upward	Superior rectus and inferior oblique
Downward	Inferior rectus and superior oblique
Inward	Internal rectus and superior and inferior oblique
Outward	External rectus and superior and inferior oblique
Upward and Inward	Superior rectus, external rectus, and inferior oblique
Upward and Outward	Superior rectus, external rectus, and inferior oblique
Downward and Inward	Inferior rectus, internal rectus and superior oblique
Downward and Outward	Inferior rectus, external rectus and superior oblique

THE REFRACTIVE MEDIA.

In order that an eye may see distinctly it is necessary that the vibrations of light that come from different distances outside of the eye be enabled to reach the inside coat called the **retina**. In doing this they pass through the transparent portions viz.: **cornea, aqueous humor, lens, vitreous humor**. All of these together act as one piece of mechanism and are called the **refractive media** because the word **refraction** means to change and adjust rays of light from one direction to another, and so these four parts act as the medium for properly adjusting the forms of light that enter the eye.

After extended study of what are considered to be normal eyes a certain positive "valuation of adjustment" has been given to this refractive media in terms of **dioptres** and is called the **dioptric power of the eye.** A Dioptre is the unit of measurement for optical lenses. 1D.=a focus of parallel rays at 1 metre from the lens; 2D.=½M. focus. Now looking at it another way we would say a lens that focuses at 1M. is a 1D. lens; at ½M. a 2 D. lens; consequently after this manner the dioptric power of the eye was figured out.

Tscherning in his "Physiologic Optics," page 31, gives the dioptric value of the **complete optic system of the eye** to be 58.38.

The cornea about 42D. and the lens 16D. The aqueous and vitreous humors having but little value in the sum total. Authors differ but little on these points so it is quite safe to say that about 60D. is the dioptric power of the fully developed eye.

A STUDY OF ACCOMMODATION.

Accommodation, in the study of the eye, means in effect, a change in the arrangement of the rays of light after entering the eye, so that whether close to, or at some distance away from the eye, the object looked at must be kept "focused" or sharply defined upon the retina. This change takes place only in the lens, not by sliding backward and forward as in adjusting a telescope; but merely by changing the adjustment of the lens from its least convexity and in this way increasing or decreasing its dioptric power.

In the study of **O**phthalmic **O**ptics and the practice of **O**ptometry it becomes necessary for the student to realize the importance of this subject.

Two special points are always to be considered:

(1) Conjugate foci; (2) Amplitude and range of accommodation.

A great deal of study has been given in an experimental way as to just how the "Act of Accommodation" is accomplished and the most satisfactory and acceptable action of the eye is as follows:

(1) **Parts concerned are**—The lens and the suspensory ligament and ciliary muscle which directly surrounds it.

(2) **Action**—The contraction of the ciliary muscle narrows the little space around the edge of the lens which has been held taut by the suspensory ligament which is attached to it, thus releasing the tension on the lens, which being somewhat elastic, increases in convexity according to the necessity of regulating the light so that it focuses on the retina properly. The relative distance the eye is from the object desired to be seen is the governing influence impelled by the brain to adjust it for that particular point.

According to scientific tests of many thousands of human eyes as regards vision it is a fact that when one is twenty feet or more away from any object he is looking at that no accommodation is necessary at any age, in the perfect eye, in order to see plainly. Here the eye is said to be at rest— meaning no accommodation or eye strain. According to the "laws of light," however, at any age, the adjustment (accommodation) becomes necessary when looking at an object at any point closer than twenty feet. The closer the object the greater the demand for the adjustment. This is what is termed conjugate foci in the sense that some one point outside of the eye is always in direct focus with the retina.

Range of Accommodation

We present here a table having reference to the fact that "Accommodation," or the power to adjust the lens, decreases gradually as the years pass.

Year	Amplitude in Dioptres	Year	Amplitude in Dioptres
10	14	45	3.5
15	12	50	2.5
20	10	55	1.75
25	8.5	60	1
30	7	65	0.75
35	5.5	70	0.25
40	4.5	75	0.00

Now upon reference to the table it will be seen that as the lens becomes harder and less elastic by age, it eventually entirely loses this power and needs artificial help in the form of glasses that will supply the deficiency. In early youth, then, we find that the range or adjustment of accommodation is the greatest, and that is why glasses become necessary for easy close work at about forty-five years of age, and thereafter an occasional change to a stronger focus is needed to keep pace with the gradual loss within the eye.

SHOWING CHANGES IN ACCOMMODATION

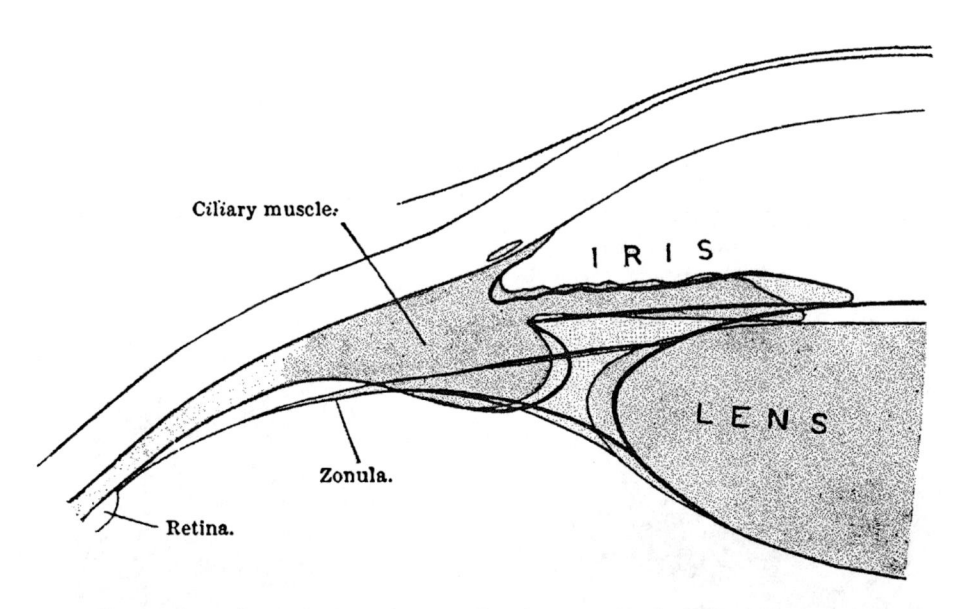

Even though one may know the anatomy of the parts involved in accommodation, still it is sometimes difficult to grasp just what does take place. The two diagrams here shown will serve to make it more clearly understood. Remember, the diagrams are flat views and the student must always have in mind that he is facing the front of the eye; that the ciliary muscle and suspensory ligament surround the edge of the lens; that the lens is at its least convexity as shown on the shaded part of the one picture. Now, when the ciliary muscle contracts it draws closer to the edge of the lens all around it equally. The lens then becomes thicker through its antero-posterior diameter. At the same time the pupil contracts a little to sharpen vision. A branch of the third nerve affecting both the ciliary muscle and the sphincter muscle of the iris act at the same time. Such a change is constantly going on as a person changes his view from one point to another.

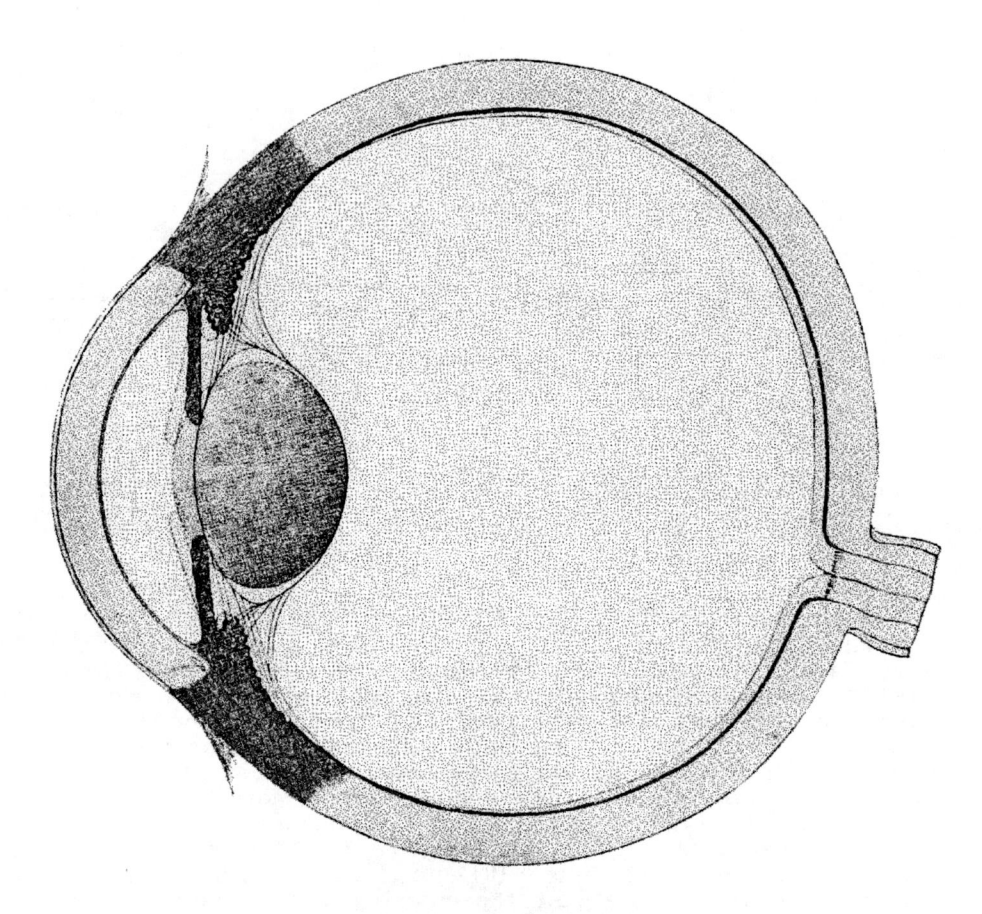

A Study of Accommodation.

Change in the curvature of the lens in accommodation according to the theory of Helmholtz.—(Modified from Landolt.)

SPASM OF ACCOMMODATION.

This term represents the "live wire" of the majority of complaints that are classed under the condition called "eyestrain." It is the fighting line between the oculist and the optometrist. It is the home office of trouble for the refractionist who doesn't thoroughly understand its little game of deception. It is the thing that really put optometry on the map where it is today. It is the thing that demands lots of respect and attention. Study it.

Getting down to facts. Spasm of accommodation means a tired ciliary muscle resulting from an excessive demand upon it to adjust and maintain more perfect vision which it becomes necessary to do when some departure from normal vision exists in the eye. A tired muscle cramps or contracts. When the ciliary muscle is tired it manifests the fact by causing distress in various ways. Knowing that contraction of this muscle is what adjusts the focal power of the lens, the student will at once realize that the nerve force used is compelled to act beyond its normal capacity and must finally become more or less inefficient. Such is the case with varying symptoms of this disorder. It being partial at times called **clonic spasm** and again more or less permanent called **tonic spasm.** When the oculist finds such a condition apparently manifest he uses "drops" called a Cycloplegic which releases the cramped condition of the ciliary muscle and enables him to get the exact refraction of the eye very readily. Being a physician he is legally entitled to use drugs according to his best judgment. The use of such drops has its inconveniences and draw-backs. Necessity demanded a different procedure that ultimately would attain the same results. After years of experimenting with Ophthalmic lenses a very satisfactory method has been developed called the **fogging system,** also an entirely different method called **static** and **dynamic skiametry** or **retinoscopy.** It is by the use of these two methods that the optometrist is able to compete with the oculist and satisfy his patients.

PART THREE

presents in simple form the sort of pathological conditions of the eye that any refractionist should be familiar with. The Optometrist has his limitations and should absolutely know them. His personal welfare and that of his patient must at all times be reckoned with. Let your motto be "Play Safe."

Now that you have become familiar with the anatomy, physiology and optics of the eye it will prove an easy matter to acquire and hold a good working knowledge of diseased conditions.

DISEASES OF THE EYE THE OPTOMETRIST SHOULD RECOGNIZE

Every refractionist before taking the first step toward the regular examination for glasses should look carefully for any unusual sign or symptom of the eye bearing upon anv past or present condition that would be likely to interfere with successful completion of his work. During the inspection he should ask the patient if at any time he has had any diseased condition of the eyes of a serious nature, as there might be some internal disease that he should know about.

Any acute inflammatory condition, sometimes even with apparently trivial symptoms may cause photophobia and ciliary spasm and interfere materially with exact work.

The list given here is merely intended as a synopsis that will convey a quick understanding of the conditions mentioned and lead to a text book on Eye Diseases for a more complete knowledge.

EXTERNAL DISEASES

CONJUNCTIVITIS—The **palpebral conjunctiva** lines the lids back to the fornix where it turns back upon the ball and becomes the **ocular** conjunctiva from there forward to the margin of the cornea.

Of course in all acute stages of diseased conditions one can refract only those of more or less mild form, if at all.

Simple conjunctivitis is merely an irritable conjunctiva which occurs from many causes including eye strain. There is no discharge of pus, but more or less increase of redness especially of the inside of the lids. Photophobia. Even if you fit glasses under such conditions the patient may return to say they are not satisfactory, when the whole complaint would really be caused by the condition of the lids.

TRACHOMA or GRANULATED LIDS.—This is chronic inflammation of the conjunctiva—always photophobia—and nearly alwavs haziness of the upper third of the cornea, called Pannus. You cannot refract satisfactorily.

PTERYGIUM—is a fan-shaped growth of the conjunctiva extending from the inner canthus to the edge and sometimes upon the nasal side of the cornea. It interferes with satisfactory refraction and should be removed by operation.

PINGUECULA—is a small, fatty deposit in the sclera between the cornea and the inner canthus. It is not a disease, no harm comes from it, leave it alone.

CHALAZION—Sub-acute tumor of one or more of the Meibomian Glands in either the upper or lower lid. No pain, usually grows larger in time and interferes with good visual acuity by pressing on the eye-ball and distorting everything seen. Cannot refract very satisfactorily on that account. Advise operation.

BLEPHARITIS—Thickening of the edges of the lids by inflammatory process or eye strain. Small scales at the roots of the cilia and sometimes pimples, patients nearly always astigmatic. Refraction not always satisfactory until cured by treatment.

HORDEOLUM—Common stye, very painful, often the result of eye strain.

ENTROPION—Edge of the lid turns in—is the result of injury of chronic disease of the lids causing the cilia to rub against the eye-ball.

ECTROPION—same cause as Entropion—edge of lid turns away from the eye-ball. Usually the lower lid. This condition results in epiphora.

EPIPHORA—is an overflow of tears upon the cheek because they cannot escape through the puncta into the lachrymal canal on account of obstruction.

LACHRYMATION—is a term used to denote an excessive flow of tears from emotional causes. No obstruction.

NEBULA—is an almost imperceptible haziness of all or a small part of the cornea.

MACULA—is a small spot or opacity of the cornea, usually of the two anterior layers.

LEUCOMA—is a dense opacity of the cornea in part or in whole and usually the result of a serious injury or disease that affects the true corneal layer

PANNUS—is a well defined haziness usually found in the upper third of the cornea, usually the result of Trachoma.

FOREIGN BODY IN THE EYE. (abbreviated F. B.)

This means anything at all that finds its way between the lids and remains there, whether loose or attached.

Where this directly concerns the Optometrist is, that it occurs very frequently that a person will call upon a refractionist anticipating relief from some recent eye trouble and demanding glasses for it. Inquiry discloses the fact that within the past few days more or less irritation with some pain has developed in one eye only. A well informed refractionist must at once conclude that it is not a case of eye strain. Upon careful inspection in such cases under a good light by oblique illumination a minute spot will be seen on the cornea, that does not belong there. If gray in appearance it is likely to be a small ulcer. If dark it is without doubt some small particle that has become imbedded in the anterior layers of the cornea and should be removed. In either case don't touch it as it is a case for the medical doctor. It is just as liable to occur soon after you have fitted that person with glasses and if so you will be the first person thought of and the blame given to you, so be careful to watch out accordingly.

CORNEAL ULCER.

Very painful, photophobia intense, lachrymation profuse, palpebral and ocular conjunctiva inflamed. Inspection will show a small gray spot on the cornea. It must have immediate and skillful attention, as if in front of the pupil it may result in partial blindness in that eye.

PTOSIS—Drooping of the upper lid. Usually congenital owing to incomplete development of the levator palpebrarum muscle. Operation does no good. If acquired, it is usually the result of acquired syphilis and means a partial paralysis of the third cranial nerve. Consequently all that the third nerve supplies is affected and we have cycloplegia-mydriasis with the cornea turned down and toward the outer canthus owing to the muscles being unable to hold it in the primary position and leaving it under the control of the external rectus and superior oblique.

ECCHYMOSIS—"Black eye," result of injury.

TRICHIASIS—"Wild Hairs" or eyelashes usually turning in and rubbing upon the eye-ball causing much distress, operation necessary for relief.

INTERNAL DISEASES

IRITIS.

Acute Iritis is very painful. Four principal symptoms are: pain, contracted pupil, iris looks dull, redness on the sclera around the cornea. Usually caused by syphilis or rheumatism. If not promptly and properly treated and the pupil widely dilated, the posterior surface of the iris becomes attached to the lens capsule. Once such a condition is established called **posterior synechia** the result is that the pupil will not react to light and also the lens has lost its adjustment for accommodation and becomes static. In such a condition it is out of the question to satisfactorily refract such an eye on account of having no way to adjust the focus. The way to detect the extent of the adhesions is to have a physician use a mydriatic.

Anterior synechia is a term applied where the front part of the iris has become attached to some part of the inside surface of the cornea, the result of disease or injury. Such a condition can plainly be seen.

OPTIC ATROPHY.

The subjective symptoms are reduction in the acuteness of vision both as to color and form, with more or less dilation of the pupil—(mydriasis). Complete blindness is the usual result of the progress of this disease. Having studied and become familiar with the appearance of the optic disc in health the examiner will quickly notice the loss of its pinkish zone as well as its minute vessels which have disappeared leaving the entire disc presenting a dull white appearance, while the blood vessels, especially the arteries of the retina, are much smaller than usual. The ball retains its normal tension and the refractive media clear. It is by this comparison that it is easy to distinguish between glaucoma and optic atrophy. It chiefly occurs in middle life and there is really no successful treatment.

EVERY REFRACTIONIST SHOULD KNOW SOMETHING ABOUT CATARACT

CATARACT is any complete or partial opacity of the lens or its capsule. There are three general terms that cover all conditions:
Congenital,
Traumatic,
Senile.

The term "congenital" implies present at birth. In many children directly after birth is found more or less opacity of the lens which condition will remain stationary throughout the life of that person. A slight opacity admits light into the eye and the actions of a child thus afflicted simulates myopia. The only remedy is surgical.

TRAUMATIC CATARACT. The term "traumatism" means injury. Anyone at any age can be thus afflicted. A blow directly upon the eye-ball will cause it. If the capsule is not ruptured it will become a permanent opacity. If, however, a small rupture of the capsule occurs permitting the lens substance to come into contact with the aqueous humor, the latter gradually absorbs it, the debris being carried off through Schlemm's canal.

SENILE CATARACT is comparatively common and likely to develop in anyone. It usually appears after the age of fifty. The real and direct cause in any given case is unknown other than we know that some interference has taken place with the nutrition of the lens usually supplied by the ciliary processes and the lymph in Petit's canal. Some cases are traceable directly to some general disease such as Diabetes, Bright's Disease of the kidneys, Arterial disease, etc.

Symptoms.—There is no pain nor inflammatory condition present. The first sign is usually diminished acuity of vision. The patient complains of seeing spots on the object looked at. The interference with vision gradually increases until finally there is only mere perception of light. In almost every instance only one eye is affected at first and progresses to quite an advanced stage before the other eye shows any symptoms whatever. It is almost inevitable, however, that the fellow eye will follow the same course in due time. The time required for full development is very uncertain. It may be very slow or may ripen completely within a few months, or it may, at a certain stage become stationary.

There are four stages of development in senile cataract:
Incipient,
Maturing stage,
Mature or ripe,
Hyper-mature or over ripe.

The Incipient or beginning condition is a more or less nebulous—slightly opaque—dull appearance of the lens in which the patient feels rather than sees there is something wrong. With a good light reflected by a retinoscope into the eye it can be detected by the observer, especially when compared with the "reflex" of the other eye. Sometimes this slightly opaque condition remains stationary for years, with comparatively little loss of useful vision. Hence it is often wise not to alarm the patient about it, but for your own protection an interested relative should be informed accordingly.

The **maturing stage** comes next.

The vision is becoming noticeably less acute as the opacity increases. Swelling of the lens increases owing to absorbing fluid between its layers. The patient, at this stage, requires less plus for reading due to increased convexity of the lens of the eye. The condition is popularly known as "second sight" and sometimes, in quite an elderly person especially, remains stationary for the balance of his life.

The **third stage** means ripe cataract. The eye has become blind owing to complete opacity of the lens, and its appearance now is a dull gray or slightly amber color. It has lost the fluid previously absorbed. Its many layers have become firmly adhered to one another and it is at this period that it is most easily separated from its capsule and for best results should be removed without much delay as further changes are likely to develop into the **hyper-mature** stage, and an operation upon over-ripe cataract is less favorable and more difficult than during the mature period.

AN APHAKIAL EYE

is an eye without a lens, usually the result of an injury or operation. Such a condition can usually be detected by noticing that the pupil dips backward instead of forward and the iris is tremulous or shakes as the eye is moved about, owing to the support of the lens which was directly behind and in contact with it having been taken away.

The usual spherical lens to correct infinity for an aphakial eye is about a plus 10. Usually a plus cylinder against the rule is required also. By no means is it usual to expect good vision in such cases and it is considered excellent results if fifty per cent vision is regained after a cataract operation. About plus three added to the distance lenses is required for close work.

GLAUCOMA

This is an important and very serious disease of the eye that every refractionist should be on the lookout for, especially in people somewhat advanced in years. When an eye is once afflicted with this disease very little can be done for it in the way of permanent relief. I can merely refer to it in a general way and strongly urge the student to study it carefully in some text book on diseases of the eye. The reason why one should know how to detect it, is that in the majority of cases in which it is developing it is found that the heretofore acuity of vision of the patient is gradually becoming less and he comes to you with all confidence expecting relief from glasses. A relief which, properly speaking, you cannot give him. It ultimately means complete blindness with no probable hope for a cure. If so, then can you not do as much for the patient, with glasses, as can the Oculist? You can, of course. But the Optometrist must always bear in mind that he is not to be considered as "the court of last resort" in a legal or properly qualified sense and in no case should assume any responsibility for suspected pathological conditions.

One special symptom to become familiar with is the "tension" of the eye-ball. First learn, by lightly pressing with the forefinger of each hand upon the healthy eye-ball, its "give and take" feeling. Glaucoma being a condition where the drainage of the debris from the eye through the spaces of Fontana and canal of Schlemm has become retarded, the eye-ball gradually becomes more tense or hard until finally it is a very easy matter to detect that fact by palpation with the finger tips.

FLOATING SPOTS IN THE EYE

Very often the refractionist will have patients who complain of the fact that "every once in awhile I have little spots, like shadows, in my eyes and they appear to move around when I move my eyes, but settle down and are quiet when I am reading or writing. There isn't any pain about it but they annoy me and I would like to know what is wrong." This condition is known as

Muscae volitantes or floating particles in the vitreous.

You will state to the patient "that in most cases they do not mean any harm as far as disease is concerned; but are usually the result of eye-strain, insomnia, indigestion, etc. All of which must be looked after and remedied accordingly." In myopia of high degree, floating specks are almost constant and are not always relieved by wearing glasses.

Having now given a general description of the principal external and internal diseased condition of the eyes that it is the duty of the Optometrist to recognize in order to "play safe" both to himself and his confiding patient, we leave this thought with him: That no person is entirely his own patient who in any way is afflicted with even the slightest pathological disturbance. An apparently simple symptom might and often does lead to serious results if not promptly recognized and cared for. A careless diagnosis with an ignorant prognosis may lead to trouble. Do not often advise nor assume any responsibility. REFER THE CASE AT ONCE TO THE OPHTHALMOLOGIST.

NYSTAGMUS

Occasionally someone will call upon you to see if you can benefit their vision with glasses. Upon the usual inspection (always necessary before proceeding with the Optical examination) you discover a peculiar and constant lateral twitching of both eyes. You have a case of genuine nystagmus. There are varieties of the movements classified under this head; but generally we find the movements or twitching of the eyeballs are rhythmic bilateral and from side to side; both to the right, then to the left, and so on, averaging in speed from one to three times in a second and to the extent of about two or three mm. to either side of the primary position straight ahead.

This condition is usually congenital and with an obscure etiology (cause.) Sometimes it is the result of some serious effection of the eyes soon after birth, resulting in corneal scars that prevent the development of good vision. Congenital cataract is also a contributing feature. When, however, the refractive media is clear the condition of suspended development is rather difficult to discover. Anyway, it's not your case, because of the fact that it has been found that glasses offer very little help in the way of improving vision. True nystagmus is not due to Optical defect and is not traceable to occupation. No perfectly satisfactory explanation of nystagmus has yet been given, other than it is a perversion of the centres for parallelism and not with the muscles themselves.

AMBLYOPIA

Strictly speaking, amblyopia is not disease in any form. We classify it here because the text books do not explain the term in a way to make it thoroughly understood by the non-medical rafractionist. Its real meaning is as follows, viz.: Diminished visual acuity, congenital, with no possible remedy. The eye is not blind nor diseased in any form. In no sense is true amblyopia an acquired condition. The refractive media is clear and may or may not be ametropic. Notwithstanding correct retinoscopic findings, the glasses do not materially improve vision then or thereafter. The ametropia in an amblyopic eye may be exactly the same as in the fellow eye which sees perfectly with its correction by glasses, while the former will not. The Ophthalmoscope or any other kind of an objective examination shows nothing wrong. What is the answer? Simply this: At one or more points from and including the retina to and including the optic tract there is an interference with the proper vibrations of light that have reached the retina, and an undeveloped condition of some unknown kind exists that obtunds detail in objects and gives only a gross image in return. If for any reason the development of rods and cones does not continue after birth the vision remains accordingly. Microscopical study of the retina shows about 3,500,000 cones in the retina at birth; and in the fully developed eye about 7,000,000. In the macula alone, a space of less than 2 mm. in diameter, there are in the developed eye 13,000 cones. How plainly then, is the fact that in any condition where the cones are less than the amount required for good vision, the eye cannot be made subject to decided improvement. Also it must be remembered that the optic tract represents a "cultivated area" that is developed only in accordance with the demand made upon it through the refractive media.

A diagnosis of amblyopia is made only by "exclusion;" meaning a thorough familiarity of all the optical, physiological and pathological conditions of the eye, and after carefully examining for all and eliminating them from the case we have only one probable condition left that in any way answers, so it **must be amblyopia.** There are other conditions of diminished vision that simulate true amblyopia that in our examination we find are false. They are classed as follows, viz.:

1. Amblyopia Exanopsia.
2. Amblyopia Toxic.
3. Amblyopia Hysterical.

No. 1 is diminished visual acuity, the direct result of uncorrected ametropia; and owing to want of optic tract training, does not immediately and fully respond to the correct glasses. If, however, the glasses are worn continually for some time there is a gradual improvement in sight until after a time it becomes comparatively normal. The history of the case differs somewhat from true amblyopia as the element of a high ametropia is always present, and the vision improves with glasses while in the true condition it does not. The point is, be guarded in your prognosis.

Toxic amblyopia is diminished vision always in both eyes, the result of auto-intoxication of some form. It may be from over indulgence in food, liquors, or drugs. Easily diagnosed; and the remedy is to cleanse the system and put it into a healthy condition after which the eyes will resume the same vision as before.

Hysterical amblyopia is practically nothing at all the matter with the eyes. It is regarded by many as a sex problem and treatment is directed toward the general nervous system. In some cases it is of only short duration although it may continue for several weeks. Judicious questions to the patient will soon bring out the true condition. Leave it alone.

PATHOLOGICAL VARIATIONS OF THE PUPIL

In every case, before proceeding with an examination for glasses the eyes should be carefully inspected for signs of abnormal conditions. This especially applies to the pupil as more often than suspected it offers a very grave prognosis at a time when the affected person is little aware that anything is seriously at fault with him.

Light reflex of the pupil means that under ordinary conditions the pupil will contract and dilate according to the degree of light to which the eye is exposed. Towards a bright light it should contract and on turning away dilate more or less. A fixed pupil never occurs in healthy individuals with healthy eyes.

While over a dozen different terms are required in explanation of pupil reflexes, those given here are the principal ones·

1. Loss of Pupillary light reflex with retention of the convergence and accommodation. (This is Argyll-Robertson pupil.)

2. Loss of convergence and accommodation and retention of light reflex (just opposite to the Argyll-Robertson pupil)

3. Loss of pupillary reflex for light, also convergence and accommodation (all three affected).

4. Abnormal miosis (contraction of the pupil) with retention of light reflex and convergence. The miosis being caused either from abnormal stimulation of the sphincter pupillae, or from paralysis of the dilator pupillae.

5. Abnormal mydriasis (dilation) with retention of convergence and light reflex. Stimulation of the dilator pupillae.

6. Anisocoria (difference in size of pupil).

7. Irregular form of pupils.

The one of special interest to the refractionist is the **Argyll**-Robertson pupil (study it in the text books).

Diagnosis.—Loss of the pupillary light reflex, with either contraction or medium dilation. Both eyes are similarly affected in the greater number of cases, although, in some it is unilateral for a long time. Again—the light reflex may be utterly lost in one eye, and only partly lost in the other. This light reflex is as a rule permanent, since it occurs the greater number of times as an accompaniment of diseases of the nervous system that are progressive, and it must be distinctly remembered that it never occurs in healthy individuals. The pupil is always uniform and should never be confounded with posterior synechia in any form, the latter being an attachment of the iris to the lens capsule resulting from iritis.

Etiology.—While the general scientific opinion is that quite all cases of Argyll-Robertson pupil are due to syphilis, there are probable exceptions in a small per cent, but it is enough to say that a most important symptom has been recognized that leads to a grave

Prognosis.—Regardless of the apparent health of the patient at the time of the examination. The fact that in this condition the accommodation is not interfered with there will be no trouble in refracting the eyes either for distant or near point glasses, resulting in reasonable satisfaction.

OPTICAL NOMENCLATURE

(Reprint from the Keystone Magazine of Optometry.)

Address by Professor Frederick Before the Annual Convention of the Michigan Association of Optometrists at Jackson, Mich.

Every profession that treats on scientific problems makes use of (glossaries) scientific technical names, for the purpose to perfect a language that may be universally understood. Optometry has such a language made up of words, roots, prefixes and affixes which are derived from the Greek or Latin, and the student who makes these derivatives a study can, with ease, unravel the meaning of words or combination of words, it matters not how complex or difficult they may appear.

For example, for "a-chroma-opsia," we look over the lesson sheet, first column, for "a," which signifies "lacking." Then look for "chroma," which signifies color; then look for "opsia," which signifies vision; therefore, the term "a-chromatopsia," meaning lacking color vision, in other words, color blind.

Again, for the word an-irida we find "an" signifies lacking, "irida" signifies iris—lacking iris. Let us now unravel the combination of an-iso-coria: "An," lacking; "iso," equal; "coria," pupil. Lacking equal pupil. Anti-metr-opia: "Anti," opposite; "metr," measure; "opia," error of refraction. Therefore, antimetropia, opposite measure of refraction, meaning one eye nearsighted, the other farsighted.

WORDS, ROOTS, PREFIXES, AND AFFIXES

Words	From	Meaning
A, An,		Prefixes, meaning "Lacking". (A-blepsia.) (An-iso-metr-opia.)
Aden	Greek	Glands, such as tear glands. (Dacry-adenitis.)
Amauro	Greek	To make dark (blind). (Amauro-sis.)
Ambly	Greek	Dull, dim. (Ambly-opia.)
Anti	Greek	Opposite. (Anti-metr-opia.)
Aqueous	Latin	Water. (Aqueous Humor.)
Argo	Greek	Non-use, out of use. (Arg-ambly-opia.)

Words	From	Meaning
Amplitude	Latin	Fullness, completeness. (Amplitude of Accommodation.)
Algia	Greek	Pain or ache. (Neur-algia.)
Ab	Latin	Prefix meaning "Away" or "From". (Adduction.)
Ad	Latin	Prefix meaning "To" or "Toward". (Adduction.)
Blepharo	Greek	Eyelid. (Blephar-itis.)
Blenno	Greek	Mucus or pus. (Blenno-rrhea.)
Brach	Greek	Short, near. (Brach-metr-opia.)
Blepsia	Greek	Vision. (A-blepsia.)
Chroma	Greek	Color. (A-chromat-opsia.)
Core	Greek	Pupil. (An-iso-corsia.)
Canthus	Greek	Corner of the eye. (Inner Canthus.)
Cata	Greek	Down or beneath. (Cata-phoria.)
Cerat / Kerat	Greek	Cornea. (Ceratometer.) Proper Horn.
Con	Latin	Together with. (Convergence.)
Copo	Greek	Strain, fatigue, worn out. (Copi-opia.)
Cyst	Greek	A sac. (Dacry-o-cyst-itis.)
Conus	Greek	A cone. (Conic Cornea.)
Contra	Latin	Opposite. (Contra generic.)
Clonic	Greek	Pertaining to confused motion. (Clonic Spasm.)
Disc	Greek	A round body. (Optic disc.)
Dys	Greek	A prefix meaning "Bad". (Dys-opsia.)
Dexter	Latin	The right. (Oculus Dexter.)
Dioptrics	Greek	Science of refraction of light.
Diopter	Greek	A meter focus as used in Optometry.
Dacry	Greek	Tear. (Dacry-ops.)
Edema	Greek	A swelling.
Erration	Latin	Wandering. (Aberration.)
Erythro	Greek	Red. (Erythr-opsia.)
Ec, or Ek	Greek	Out. (Ec-tomy.)
Ectos	Greek	Outside.
Eso	Greek	In. (Eso-phoria.)
Em	Greek	A prefix meaning "In". (Em-metr-opia.)
Exo	Greek	Out. (Exo-tropia.)
Focus	Latin	Coming to a point.
Fundus	Latin	Foundation.
Graphy	Greek	Description, drawing.
Hydro	Greek	Water.
Hemi	Greek	Half. (Hemi-an-opsia.)
Hetero	Greek	Others, or many. (Hetero-phoria.)
Hyper	Greek	Over, above. (Hyper-phoria.)
Hypo	Greek	Under, or beneath (Hypo-metr-opia.) (My-opia.)

Words	From	Meaning
Itis	Greek	Inflammation. (Retin-itis.)
Iso	Greek	Equal. (Iso-coria.)
Image	Latin	Resemblance.
Irid	Latin	Iris. Rainbow. (Irid-o-plegia.)
Kinetic	Greek	Motion, to move.
Leuko	Greek	White. (Leuk-oma.)
Lachrymal	Latin	Pertaining to tears. (Lachrymal ducts.)
Meter	Greek	Meter or measure. (A-metropia.)
Micro	Greek	Small. (Micr-opsia.)
Megalo	Greek	Large. (Megal-scope.)
Mono	Greek	One or single. (Mon-ocular.)
Nepho	Greek	Cloud. (Neph-a-blepsia.)
Neuro	Greek	Nerve. (Neur-a-sthenia.)
Nyct	Greek	Night. (Nyctal-opia.)
Nebula	Latin	Cloud, mist.
Opia, Opsia, Ops.		Affixes.
Opia	Greek	When applied to a word or a combination of words, gives it the meaning of "An error of refraction vision which may be corrected by glasses", as Hyperopia, Myopia, Presbyopia.
Opsia	Greek	When applied to a word means "Vision". (Chromat-opsia.)
Ops	Greek	When applied to a word means "The eye" (Megal-ops.)
Ortho	Greek	Straight, normal. (Orthoporia) normal fixation.
Oculus	Latin	Eye.
Oma	Greek	A growth or new growth. (Trach-oma.)
Photo	Greek	Light. (Photo-phobia.)
Phakia (phacia)	Greek	Lens. (Crystalline lens.) (A-phakia.)
Plegia	Greek	Stroke. (Paralysis.) (Cyclo-plegia.)
Ptosis	Greek	Falling. (Blephar-ptosis.) (Phacia-ptosis.)
Phobia	Greek	Fear. (Photo-phobia.)
Punctus	Latin	A point. (Punctum Proximum.)
Pseudo	Greek	False. (Pseudo Myopia.)
Phoria	Greek	Tendency to swerve. (Hyper-phoria.)
Pyo		A prefix denoting "Pus". (Pyo-rrhea.)
Poly	Greek	A prefix denoting "Many". (Poly-coria.)
Palpebral	Latin	Pertaining to the eyelids.
Presby	Greek	Old. (Presby-o-pia.)
Rhea	Greek	A flow or discharge. (Dacry-o-rrhea.)
Retina	Latin	A network of sensitive layers.
Spectrum	Latin	An appearance. (Solo-spectrum.)
Spasm	Greek	Contraction, cramp. (Clonic spasm.)
Senilis	Latin	Pertaining to old age. (Arcus senilis.)
Sthenic	Greek	Strong, or strength. (A-sthen-opia.)

Words	From	Meaning
Stigma	Greek	A point. (A-stig-ma-tism.)
Sclera	Greek	Hard. (Scler-o-sis.)
Super	Latin	Prefix meaning "Above" or implying excess. (Superior.)
Staphy	Greek	A grape. (Staphyl-oma.)
Scope	Greek	Observation, to view, viewing. (Telescope.)
Synechia	Greek	Adhesion. (Blepharo-Synechia.)
Sinister	Latin	The left. (Oculus Sinister.)
Sis	Greek	An Affix. Action.
Toxin	Greek	Poison. (Toxic Ambly-opia.)
Tomy	Greek	Cutting. (Irid-ec-tomy.)
Tome	Greek	A cutting instrument.
Tropia	Greek	Turning. (Exo-tropia.)
Tonic	Greek	Continued. (Tonic Spasm.)
Umbra	Latin	Shadow. (Umbra-scope.)
Ula	Latin	Small. (Lentic-ula.)
Vergence	Latin	Inclination. (Convergence.)

CPSIA information can be obtained
at www.ICGtesting.com
Printed in the USA
FSOW02n2037250516
20821FS